Canned Laughter

Canned Laughter

The Best Stories from Radio and Television

PETER HAY

New York   Oxford
OXFORD UNIVERSITY PRESS
1992

Oxford University Press

Oxford   New York   Toronto
Delhi   Bombay   Calcutta   Madras   Karachi
Kuala Lumpur   Singapore   Hong Kong   Tokyo
Nairobi   Dar es Salaam   Cape Town
Melbourne   Auckland   Madrid

and associated companies in
Berlin   Ibadan

Copyright © 1992 by Peter Hay

Published by Oxford University Press, Inc.,
200 Madison Avenue, New York, New York 10016

Oxford is a registered trademark of Oxford University Press

Library of Congress Cataloging-in-Publication Data
Hay, Peter, 1944–
Canned laughter, the best stories from radio and
television
Peter Hay.
p.   cm.   Includes bibliographical references (p.   ) and index.
ISBN 0-19-506836-X
1. Broadcasting—Anecdotes.   I. Title.
PN1990.87.H38   1992   791.44—dc20   92-6011

Every reasonable effort has been made to contact copyright holders to secure permission to quote passages from works not in the public domain. Any omissions brought to the publisher's attention will be remedied in subsequent editions.

Grateful acknowledgment is hereby made for permission to reprint excerpts from the following works:

Chatterboxes, by Brian Johnston; reprinted by kind permission of Mr. Murray Pollinger and Methuen, London. Copyright © 1983 by Brian Johnston.

Coming to You Live!, by Denis Norden, Sybil Harper, and Norma Gilbert; reprinted by kind permission of Denis Norden; Norma Gilbert and her agent, Rivers Scott; and Methuen, London. Copyright © 1985 by Denis Norden, Sybil Harper, and Norma Gilbert.

I Looked & I Listened, by Ben Gross; reprinted by permission of Random House. Copyright © 1954, 1970 by Ben Gross.

In Town Tonight, by Peter Duncan; reprinted by permission of Random Century Group. Copyright © 1951 by Peter Duncan.

More or Less, by Kenneth More; reprinted by permission of Hodder & Stoughton. Copyright © 1978 by Kenneth More.

And So It Goes, by Linda Ellerbee; reprinted by permission of the Putnam Publishing Group. Copyright © 1986 by Linda Ellerbee.

Television—A Personal Report, by Robin Day; reprinted by kind permission of Sir Robin Day. Copyright © 1961 by Sir Robin Day.

Writing for Dough—Adventures of a T.V. Comedy Writer, by Bill Idelson; reprinted by kind permission of Mr. Bill Idelson. Copyright © 1989 by Bill Idelson.

2 4 6 8 9 7 5 3 1

Printed in the United States of America
on acid-free paper

Acknowledgments

With this my ninth book, I am happy to round up the usual suspects whose professional advice or personal friendship made this book possible.

As always, it has been a privilege to work with the superb team led by President Ed Barry and my patrician editor Sheldon Meyer at Oxford University Press in New York, and especially Leona Capeless, Stephanie Sakson-Ford, Scott Lenz, Joellyn Ausanka, Laura Brown, Vera Plummer, Jonathan Weiss, Linda Robbins, Ellen Chodosh, and many others who produce and promote my books. Last year in England, I was delighted to meet also some of the good people at Clarendon Press: Juliet New, Coleen Roan, Richard Sansom and Simon Wratten.

My gratitude to the many broadcasters on whose programs I have appeared and who contributed to this book or helped to promote previous ones: especially Stuart Allingham, Jan Black, Michael Cart, Paul Chambers, Bonnie Churchill, John Clemens, Jim Eason, Alan Farley, Ira Fistell, Sonya Friedman, Vicky Gaboriau, Miklós Györffy, Joseph Hurley, Ellen James, Guy LeBow, Leonard Maltin, Larry Mantel, Connie Martinson, George Putnam, Tom Snyder, Siobhan Synnot, Jan Wahl, and Paul Wallach.

Among the producers, personalities, and fellow fighters I recall fondly from a decade of free-lancing in Canadian broadcasting: Harry Boyle, Neil Dainard, Patrick Gossage, Herschel Hardin, Don Harron, Hilda Mortimer, Don Mowatt, Fred Peabody, George Ryga, Pia Shandel, Robert Weaver and Jack Webster.

My heartfelt thanks to my family, especially to my wife Dorthea and my mother Eva; to my colleagues at First Stage, and to my many friends, including Laurence and David Ambrose, Dorothy Atwater, Polly and János Bak, Lia Benedetti, Alan Brock, Margaret Burk, Didi Conn, Jacqueline Crossland, Dan Davis, Michael and Mimi Donaldson, June Dragman, Linda Elstad, Martin Esslin, Jacqueline Green, Leonora and David Hays, Endre Hules, Bill Idelson, Paul Jarrico, Donna and John Juliani, Richard Kahlenberg, Patricia Keeney, Ed Kenney, Serge Leslie, Marcella Maharg, Jennifer Marcus, Ralph Maud, Virginia Morris, Eva Nemeth, Ed O'Neill, Doris Niles, Edith Oliver, Deborah Pietruska, Ethan (John) Phillips, Burt Prelutsky, Don Rubin, Catherine Rusoff, Dennis Saffren, John Sarantos, Brad Schreiber, Marsie Sharlatt, Roberta Sherry, David Shire, Eve Tettemer Siegel, James F. Skaggs, Adam Soch, Carol Sorgenfrei, Harold Spector, Loren Stephens, Sylvia Sur, Michael Szasz, Leslie Westbrook, Betty White, Carl Wilhite, Robert Windeler, John York, and Marc Zicree.

I am indebted to the Society to Preserve and Encourage Radio Drama, Variety and Comedy (SPERDVAC), to the Museum of the Moving Image in London and the Museum of Broadcasting in New York, and to knowledgeable bookmen and women in bookstores or libraries, including M. Taylor Bowie, Charles Cooper, Jerome Joseph, Gerald Kahane, Michele Merrill in the Beverly Hills Library, Anthony Anderson at USC, and to librarians at the University Research Library at UCLA, at several branches of the Los Angeles City Public Library, and especially at the Pasadena and Glendale city libraries (Bill Trzeciak has been most helpful at the latter), my two homes away from home.

For Michael Donaldson,
entertainment attorney
and an entertaining friend,
with affection, admiration, and thanks

Contents

Introduction

Norman Corwin, one of the pioneers of serious radio programming in the United States, wrote in the conclusion to his 1983 book, *Trivializing America*, about how "our culture has been reduced to vassalage within an imperious co-dominion of film, TV and sports." Corwin cites that "an actor became president; another actor is elected United States senator; an airport is named after a third actor; a highway is named for a comedian; a school is named after a first baseman; Shirley Temple grows up to be an ambassador." I don't know if these omens are any worse than highways named after bad presidents or corrupt congressmen; and I do think that Mrs. Black has performed better diplomatic service for her country than some of the businessmen who contributed to a presidential campaign, or than career civil servants with no achievements except in departmental infighting. What seems indisputable is that radio and television changed fundamentally the way we feel about actors, and some politicians. Ira Fistell, the popular talk-show host on KABC radio in Los Angeles, calls Ronald Reagan "the nicest guy in the world, but so dumb you wouldn't believe it. I know why the press was so kind to Ronald Reagan. It was unfair to criticize him. It was like taking candy from a baby." But the people who called Reagan an airhead, underestimated him. He had already visited millions of American homes as one of the good guys on the late-night show and as a trusted spokesman for General Electric, and his silken, earnest voice came through the radio every week with a little homily that a child could understand. Familiarity can also breed acceptance.

As a movie star, fixed in cold interstellar space, Reagan might not have become president; as a pitchman he did, becoming so identified with his role that he no longer cared about being an actor. Actors who survive too long on a series have difficulties escaping from their characters or finding acceptance in different roles (a subject first explored in Frank Marcus's play, *The Killing of Sister George*). Reagan was lucky in finding the role of a lifetime—which is the subtitle of Lou Cannon's book of his presidency.

When we leave home to see a play or a film, we have usually no problem distinguishing our favorite stage or movie actors from specific roles; it is the actor, not the part, that we admire. The distinction becomes blurred with radio and television, which work through personalities, whether they are actors or not. We come to feel about characters on long-running programs and soap operas the way we do about real game-show hosts and anchors, tennis players and comics, whom we see grow up and old on television. They are folks like us. When we love Lucy, we forget about Lucille Ball, who has appeared in many other roles. For years Mary Tyler Moore acted the character of Mary Richards, working in a fictional Minneapolis newsroom, but the long-running sitcom was called after her real name. The paradox was never explained. When Murphy Brown recently threw a baby shower, five real anchorwomen showed up to express solidarity; the next morning they were interviewing on their own programs Candice Bergen who would tell the newswomen that they could act better than she could. No wonder Vice President Dan Quayle has difficulty telling apart who is real.

In a world of duplicity run by lying politicians, Americans used to draw comfort from the belief that the avuncular Walter Cronkite would not lie to them. In *M*A*S*H* they found much needed catharsis from the Vietnam war, which most people watched from the living room. Here was a sitcom which made them laugh and care—at least about the actors who impersonated the medics and the soldiers. When Jane Pauley left as co-host of *The Today Show*, people cried around the breakfast tables; Little Jane was leaving home and striking out on her own. In real life, Pauley was a mother of three and pushing forty, but the fact that most of us have never met her or Uncle Walter is irrelevant. We have spent much more quality time with them than with real uncles, sisters, or, in some cases, spouses. Similarly, because of their constant, and largely agreeable, companionship, we somehow believe that these figures from our extended family know us and have watched us grow up. Television and radio have become much more to us than pieces of furniture, electronic wallpaper, or babysitters, as these media have been called. They have taken over many of the functions of family, with endless reruns providing us with albums through which we can relive the way we were and sometimes reflect on what has become of us.

On the other hand, we find it hard, as we do with family members, to forgive these figures if they do something out of sync with their on-screen personas. People were disconcerted when Arthur Godfrey suddenly fired his protégé, Julius LaRosa, on the air, and I doubt if Pee-Wee Herman will ever be truly forgiven, after the actor behind the character was arrested for indecent exposure.

Inevitably, this book of stories about radio and television reflects such paradoxes and ambiguities. Some figures who are practically family members to one reader will appear to another as complete strangers. There is a fair chance that I missed someone's favorite program or personality, simply because I was yet unborn, had tuned to another channel, or was living in a different country; none of us is local in the global village. Fortunately, I have spent roughly one quarter of my life in four different countries, so my locality and experience of broadcasting changed every dozen years or so. I missed by a couple of years Grace Archer's death on the BBC, and I did not see Jack Paar weep on NBC, but I wept, listening in my native Budapest to the final match in the 1954 soccer World Cup between the favored Hungarian team and West Germany. Hearing the game on radio—there was no television in Hungary yet—broadcast from abroad, where travel was forbidden, we were taken as a nation on a roller-coaster ride from hysterical hope to uncontrolled despair by György Szepesi, a shamelessly emotional commentator. How different that game sounded to me as an adult, in German, as the acoustic background in some scenes of Fassbinder's film, *The Marriage of Maria Braun.*

Context is everything, and, as with my other anecdote collections, I provide merely pretexts for you, the reader, to bring forth your memories and your own nostalgia. Although none of the stories that follow are personal reminiscences, the selection was inevitably inspired by my interests and limited by circumstances. In most cases, I chose story over character; the actor, not the part; the person rather than the persona. Even when the subject was fiction, I sought the underlying reality. For people who confuse anecdotes with trivia, I tried to show up the difference.

Finally, all books are limited by space and the time available to fill them. The vast majority of the stories are from the past, golden or otherwise, but I added one or two from the daily press even on the day I had to send this manuscript to the publisher. There are many more anecdotes left out than could be included, just as there are many more tapes stacked next to my VCR than extra hours to watch them.

Los Angeles, Calif.   P. H.
June 1992

CHAPTER 1

Under the Ether

Lonelyhearts

Before radio became commonplace, the mysteriousness of the voices coming through the ether into the home gave many listeners a heightened sense of intimacy. The phenomenon of listeners thinking that the voices addressed them directly, or were meant only for them, sometimes led lonely or disturbed people to complex delusions. Marcel Laporte, an early Parisian announcer who became known simply as Monsieur Radiolo, printed in 1925 some of the letters he received from listeners to his concert series.

"Monsieur," began a typical one from a woman named Suzy, "I am taking this step with a feeling of some shame. I listen to you each day. The inflections of your voice are so caressing that they produce a profound impression on my sensibility, and I feel very drawn towards you. . . . I've built a dream, monsieur Radiolo, and I don't dare to share it. In hearing your voice I've imagined myself . . . but dreams are foolish.

"Monsieur, I hardly know anything about you. Are you young, handsome, dark or blond? Perhaps one could tear away the veil of your anonymity. . . . I feel I could derive a great deal of joy from meeting you, even if we would have to maintain a banal tone in our public conversation. Would it be possible? On day X in June, I will be listening to your announcement of the evening concert. After you will be saying: 'And now we go on to our next number,' add the words: 'give me a declaration.' This will offer me the clue to your inclination to spend some time with me, and I shall write again to fix a meeting place."

When the June concert came and Radiolo did not utter the extra words, he received another letter from Suzy:

"Monsieur, I am devastated and desperate. You did not say the phrase I was hoping for. I am now banished to my sad thoughts and the sole consolation of listening to you in my living room. I had heard that your picture appeared in a magazine, and I will place it on the radio set. . . . Obviously you have a girlfriend, perhaps a married woman, to whom you recite, under conditions of intimacy, those pretty verses that we hear from afar. We are finished, I won't be writing to you any more. . . . Adieu, and have pity."

Another woman, calling herself simply "Caprice," was more practical:

"Monsieur, these words in haste are written without reflection, because if I thought too much, I would not be writing this letter. I have reserved a seat in the studio, to have the pleasure of seeing your face in real life, since I only know it from photos. Otherwise I know you by listening to the radio, and I am quite happy about that . . . very happy, in fact. Still, I would like to meet you under more intimate circumstances, so at the end of the concert, if you would be willing to spare me a few moments, please stand in front of the fashion boutique facing the artists' door to the studio. Hold a folded newspaper in your hand. I will be passing with my car and I will take you for a ride. See you soon, won't I?"

Regretfully, Radiolo remarks, he passed up the opportunity.

Familiarity

Unlike stage or movie stars, whose faces became familiar to fans, the early person-alities on radio became famous without being recognizable. The mystery enhanced their celebrity. A young unemployed singer, walking down Broadway after jury duty, wandered into WEAF, the new station owned by the American Telephone and Telegraph Company, and auditioned. Within two years, Graham McNamee was one of the best-known names in America. He announced the World Series games, told about the political conventions, and became part of people's daily lives. He himself was both amused and bewildered about it.

After those first early games when the fans learned to know by name the announcers who talk to them each night, they treated us much the same as the followers of baseball do their pitchers, often cheering us by mail or telephone, and occasionally "panning" us. Sometimes the intensity of their enthusiasm was both amusing and embarrassing. After the Los Angeles assignment, I was riding back in the smoker when I heard two radio fans discuss six- and seven-tube sets, and then fall back on the announcers.

"Who broadcast for WEAF last night?" asked one who seemed to defer to the other's vast store of information.

"McNamee," responded this mine of information: "He's good" (mentally I thanked him for that, then cocked my ears), "an old pal of mine."

"You know him?" said the first. "Where didja meet him?"

"Oh, a lotta times and a lotta places. Had dinner with him last night."

"What does he look like?"

"Sort of light-haired, 'n a big guy like Babe Ruth—and some swell dresser."

Now it happens I am dark and of medium height and my wife tells me I'm anything but a Beau Brummel; besides, on the evening in question, I had been so rushed that I had dined alone. So I thought of introducing myself, for the fun of the thing, but decided not to. I'd hate to embarrass a man like that; besides it was just another little tribute to the almost universal interest in radio.

Too Many Points of Light

In the thirties and forties, Walter Winchell had unparalleled influence on the American public both with his widely syndicated columns and with his Sunday night broadcasts. An early supporter of Franklin D. Roosevelt's New Deal, and one who recognized the dangers of European fascism to an isolationist America, Winchell willingly allowed himself to be used by FDR for causes in which they both believed. A year before Pearl Harbor, Winchell was commissioned a lieutenant commander in the U.S. Navy, and one of his jobs was to raise money for Navy Relief. He collected close to $250,000 with a single benefit at Madison Square Garden.

The Navy brass was impressed, and they next asked Winchell to appeal to private boat owners to lend their vessels to the Navy and Coast Guard to patrol U.S. territorial waters against the threat of German submarines. Payment would be a symbolic one dollar per boat. Winchell went on the air with his familiar greeting, "Mr. and Mrs. United States and All the Ships At-Sea!" He asked for any kind of skiff, motor boat, or yacht that people could spare for Uncle Sam. A few days after his broadcast, the Navy called again:

"For goodness' sakes, don't mention it again on your next broadcast. We got over two thousand boats—way too many!"

Winchell was next given the task of recruiting V-seven cadets for Naval Aviation. After three broadcasts, the Navy begged for him to stop:

"No more! We can't handle the ones we've got, thanks!"

Winchell went back to fundraising for Navy Relief, until following a presidential press conference, FDR said to him:

"Cut out the Navy Relief stuff, Walter."

"Good heavens, Mr. President, why?" Winchell asked.

"Getting too many complaints from other worthy causes, like the Red Cross. They feel the Navy Relief is getting all the money."

Supply and Demand

Around 1950, Washington broadcaster Robert Trout visited a dog pound, where he was so moved by the heartrending barking and whining of the abandoned animals that he decided to do something. He announced on his newscast that anybody who sent in five dollars to the shelter would receive a dog in return. A few days later, the director of the pound, reported to Trout about the effectiveness of his appeal:

"The five dollar bills are still pouring in. In fact, we've gotten so many of them that we have had to hire a dozen dog-catchers just to keep up with the demand!"

A Little Knowledge

Eddie Cantor's *Chase and Sanborn Hour* was the most popular variety show in the early thirties; half of America's radio households were tuned in every Sunday evening.

Jimmy Wallington, the announcer, made the mistake of mentioning that Samuel Johnson drank 24 cups of coffee a day, and the good doctor might have consumed even more if he could have bought a product from Chase and Sanborn. NBC received hundreds of calls to point out that the 18th-century man of letters drank tea, not coffee.

Forgetting just how powerful the medium had become, Eddie Cantor told his audience one night that it was his fortieth birthday, that he liked chocolate cake, and he gave his shirt and sock size on the air. He received 15,000 gift packages the following week.

Send Me Your Burnt-out Lightbulbs

George Burns and Gracie Allen did one of their routines about saving old light bulbs, which went something like this:

"George, do you have any old light bulbs?"

"I throw them away, Gracie. Why?"

"My sister could use a few."

"What for?"

"She puts them in all her lamps."

"Your sister puts old light bulbs in her lamps?"

"It's a big saving."

"A big saving?"

"Of course, George. If you put in new bulbs, they just burn out and you have to change them."

In the two weeks following the broadcast, the Burns's apartment came to be filled with old, burned-out light bulbs of various sizes, which had been mailed to them by fans.

Boomerang

Larry King, the popular American radio and television host, tells the story of Rod McKuen coming on his show once in connection with his latest book of poetry. Rather unusually, McKuen told King beforehand that he did not really want to "hype" his book during the interview. Instead, he made just one pitch at the top of the show:

"I have a new book out, and anyone who buys the book and cuts the corner flap and sends it to me will get a free copy of my last record."

That ingenious bit of promotion cost the poet two hundred and fifteen thousand records, much more than the royalties from his book.

Death in the Family

Audiences become so involved with long-running radio and television series that the death of a favorite character can cause severe and widespread trauma. An early example of such grief happened in England, on the evening of September 22, 1955, when listeners to *The Archers* were horrified by the sudden, accidental death of Grace Archer, as the character tried to rescue a horse from a blazing stable.

Anguished listeners flooded the BBC's switchboards, some crying while others asking where they could send flowers. One man called to say that his pregnant wife had collapsed in hysterics, and another that his mother had suffered a heart attack. Some people wanted to know whether it was the character or the actress who had died.

As calls and letters continued to pour in, the BBC felt it necessary to issue a statement: "She died in the normal course of duty. There has been much regret, sorrow, and some indignation." The producer of the series defended Grace Archer's demise: "I have no regrets about this. Our aim is to make *The Archers* realistic. Tragic deaths do occur in real life. If listeners are distressed by Grace's death, that is all to the good because it shows that this radio family had become real to them."

The decision to kill off Grace turned out in fact to be not so much an artistic blow for realism, but part of the BBC's campaign to fight off the new competition from commercial television. Her death, planned and executed with such secrecy that many members of the cast were not aware of it until the day of the recording, happened to coincide with the debut of Independent Television. The *Manchester Guardian* proposed the following epitaph for Grace Archer under the motto: "Dulce et decorum est pro BBC mori":

> She dwelt unseen amid the Light
>   Among the Archer clan,
> And breathed her last the very night
>   That ITV began.
>
> A maiden in a fantasy
>   All hidden from the eye.
> A spoken word; the BBC
>   Decided she must die.
>
> She was well loved and millions knew
>   That Grace had ceased to be;
> Now she is in her grave—but, Oh,
>   She scooped the ITV.

A Face That Tells the Truth

Sir Robin Day, the first newscaster on Britain's Independent Television News, which began in 1955, remembers worrying in the early days about how the average viewer received his daily dose of good and bad tidings. One morning he got a considerable lift when a middle-aged woman came up to him on the London underground train to tell him:

"Mr. Day, I *so* like the way you do your news. You don't look as if you believe a word of it."

Auntie

The BBC is known by an affectionate nickname—"Auntie"—which is appropriate enough for a rather dowdy maiden aunt. One drunk who rang the switchboard late one night became too familiar:

"I shay, isn't that old Auntie BBC?"

Eileen O'Sullivan, a much loved supervisor, happened to be on duty, and she rose to the occasion:

"Yes, this is Auntie speaking personally," she said. The voice faded into confusion.

The Art of Complaining

Because the BBC is a public corporation, it is highly sensitive to public criticism and feedback. Colin Reid was Duty Officer from the late 1960s in charge of a small staff that had to field some 60,000 calls a year: complaints ran ahead of approving comments about 8 to 1.

When in the summer of 1975 the BBC introduced live broadcasts from the House of Commons, listeners were divided. Reid recalls one woman calling up to say that she could not concentrate on ironing her laundry because of all the racket emanating from the "Monkey House"—a frequent epithet applied to the Mother of Parliaments. A listener complained that "this wretched parliamentary broadcasting constitutes a new form of mental cruelty," while another suggested that it could replace the popular quiz show, *Does the Team Think?*

Even when the listeners were not complaining, some of the approbation was definitely back-handed. Colin Reid remembers a wave of approval when a journalists' strike, in late 1975, took all news bulletins off the air.

"This is the nicest thing that has happened for a long time," one listener called in. "Delighted to be without incessant news."

"Much better to have music instead of beastly news," echoed another. And a practical lady suggested:

"If the BBC wants to save money, they have shown us a heaven-sent way of doing just that."

Keep 'Em Coming

Godfrey Talbot found reporting on the British royal family for the BBC a high-profile position. He received a great volume of mail. One enraged colonel from Cheltenham castigated him for appearing to be drunk, and demanded that he should refrain from "exalting peasants and the discourteous habit of referring to our Dear Majesty merely as 'she.'"

Among the letters he treasured was from a lady writing from Republican France, who began: "Dear Monsieur Godfrey, I am a fervent of your emissions."

The Miracle of Wireless

Godfrey Talbot also accompanied Prince Charles during a tour of Wales which followed his official investiture as the Prince of Wales in 1969. In remote villages he constantly ran into problems trying to get to a phone so that he could do live newscasts on the Royal Progress. One day, Talbot recalled in his memoirs, he dashed into the butcher shop, owned by a Mr. Jones, just in time to use his ancient telephone set and get on the one o'clock news in London. Mrs. Jones was listening to the radio in the parlor at the back of the store, when she heard the announcer introduce "a report from Godfrey Talbot direct from Merioneth."

"Now I know they don't tell us the truth," she declared. "He can't be on the wireless 'cos he's sitting next door, in our shop!"

Hi, Mom!

As more and more radio sets were sold, the fascination with the new medium grew. People did not want to be passive listeners; they sought to participate, if only they could. Graham McNamee described in 1926 how some fans tried to take to the airwaves, as they often do today.

In those early days, we had considerable trouble through thoughtless conversation from visitors who were occasionally allowed in the studios while we were "on the air," also with people who tried to sneak over messages. A broadcasting station such as ours, which is primarily designed for entertainment, is not allowed to send out personal communications; and we had to be on the alert, for every once in a while we would see someone creeping up to the microphone, just after the artist was through, and before the button was pressed, and getting over a "How are you, Mother?" or "Hello, Mary, out in Flatbush!" Of course, arrangements had been previously made with "Mother" or "Mary" to tune in at that hour. Usually they were disappointed, or else half the message got over with disastrous results; the first syllable of the "Hello" for instance, which was precisely the case one day, as I found out the next morning, when the letters came in. One thought that "Hell," which came right after a rather religious number, was not particularly in keeping. If I didn't like the singer I shouldn't swear at her. And a nice old lady, thinking it was I and not, of course, knowing it was some visitor that had been guilty, wrote that, "it was a shame, when the young man had such a pleasant voice, that he hadn't been better brought up."

Crossed Wires

During a transmitter failure in Chicago, an irate woman called up WGN and asked host Wally Phillips why he had gone off the air. The talk show host patiently explained the technical difficulties and promised that it would be soon fixed.

"Well, you certainly could have made that announcement on the air!" And she hung up.

Sally Jessy Raphael on WMCA in New York got a call one Christmas morning from a listener wanting to know why she was not attending Mass.

"I am at Mass," Raphael replied, "this program is on tape." The listener accepted the patent absurdity and hung up much relieved.

Confession

Radio is local and has an immediate impact on the audience. In the forties, a listener called in to a Hollywood station, because she missed some details of a sale that had just been mentioned on the air.

The switchboard connected her to the announcer, but he could not recall the details.

"But you just read the announcement over the air," the caller insisted.

"That's correct," the announcer replied, "but I wasn't listening."

Chapter and Worse

Ed Busch was chatting once to a woman on his show at WFAA in Dallas, when she quoted the Bible to support a point in her argument.

"Sorry, but I don't read the Bible," said the broadcaster.

"Then what the hell," the caller shot back, "are you doing on radio?"

Make My Day

Most radio talk-show hosts have a smooth, almost ingratiating manner with their public who provide them with so much free programming. Joe Pyne, who spent many years in Los Angeles, became famous in the profession for insulting his listeners. "Why don't you go gargle with razor-blades!" he told one caller. Fellow broadcaster Hilly Rose likened Pyne to comedian Don Rickles, "but Rickles is supposed to be funny. Joe Pyne took your breath away. There was nothing funny about his approach. If there was a pimple at the end of your nose, he would ask directly why you didn't cover your ugly face with a paper bag."

Rose recalls being in the studio when Pyne told an incensed caller with a heavy Yiddish accent: "Take that yarmulke off your head and shove it down your throat!"

Joe Pyne appeared on television during the height of the Watts riots. He opened a desk drawer, which had a revolver in it, and in his best Clint Eastwood manner announced: "Let them come. I'm ready for them."

Homeopathy

Jim Dunbar, a well-known liberal broadcaster in the sixties, enjoyed getting a rise out of opinionated extremists who called him on San Francisco's KGO. After listening to a familiar tirade from an old woman against communists, Dunbar asked her:

"Lady, are you aware that Lawrence Welk may be a communist?" The caller was silent. "Well, who else would put the Lenin sisters on the air?" Dunbar baited her.

Q.E.D.

When Wally Phillips launched his call-in program in the late sixties on WGN in Chicago, he settled on an opening topic. He wanted to discuss with callers public apathy. Only two people called in the entire program.

"Now that's apathy," Phillips commented afterward.

The Neighbor Lady

The most lasting impact of local radio came in small towns, especially in rural America, where communities tend to be isolated by great distances. Until her retirement in 1986, Wynn Hubler Speece, known as the "Neighbor Lady," dispensed homemaking advice, recipes, and entertainment to farmers' wives for more than four decades in over ten thousand broadcasts originating from WNAX in Yankton, South Dakota. (Earlier, the same station gave young Lawrence Welk his first radio exposure as an accordion player and later as bandleader. Tom Brokaw, NBC news anchorman, hails from nearby Webster.)

As her title suggests, the Neighbor Lady established intimate contact with thousands of her listeners, many of whom considered her a member of their family. They would often write to her and share minute details of their lives. They also told her about mishaps that took place as a result of her broadcasts. One woman wrote the Neighbor Lady from St. Peter, Minnesota, that she was listening in May 1945 to a recipe for chocolate rolls, when "your program was interrupted at Yankton telling about Hitler's death. The announcer said, 'That's all we have on Hitler's death.' Then your voice came back on, saying ' . . . mix in the flour and baking powder and bake in a hot oven ten minutes.' There were four of us here, and we all agreed that ten minutes wasn't long enough in the oven for him."

Catch That Phrase!

Much of the crazy humor Britain has exported worldwide, in *The Goon Show*, *Monty Python's Flying Circus*, and other shows, could trace its ancestry to a popular variety program called *ITMA*. The initials stood for "It's That Man Again!"—a phrase popularized by the *Daily Express* in headlines chronicling the doings and sayings of Adolf Hitler in the late thirties. The show, starring comedian Tommy Handley, began its ten-year run on the BBC in July 1939, but it was the outbreak of war, in September, that established *ITMA* as a national institution and provided a major weapon—of laughter and ridicule—in the fight against Nazism.

Cascading with word-play, street slang, and horrendous puns, which were delivered at very high speed, *ITMA* delighted through brilliant use of language, which of course has been the staple of British theatre from Shakespeare to Tom Stoppard. Here is an example of *ITMA* humor from Tommy Handley's introduction to the first show broadcast after the outbreak of war:

"Heil folks—it's Mein Kampf again—sorry, I should say: hello folks, it's that man again. That was a Goebbled version—a bit doctored. I usually go all goosey when I can't follow my proper-gender. That broadcasting ship of mine was commandeered and scuppered at Scapa and I've been taken over by the Government. Yes, they've made me Minister of Aggravation and Mysteries and put me in charge of the Office of Twerps, otherwise known as ITMA."

ITMA generated a huge number of catchphrases that seeped into English conversation. There was a German spy in the show called Fünf, who would deliver threats on the telephone.

"Otherwise sober and responsible citizens," Peter Black wrote in his informal history of the BBC, "used to telephone each other for the sheer bliss of saying 'This is Fünf speaking.' . . . The pilots of the RAF would shout, 'I'm going down now, sir!' and 'After you, Claud!' over their intercoms as they attacked. Ted [Kavanagh, the chief writer of *ITMA*] quoted a boy buried in bomb debris who patiently asks his rescuers, 'Can you do me now?' and doctors told him of patients whose last coherent sentence was a murmured 'TTFN' [= "Ta-ta for now"]. It got so that the accidental use of an *ITMA* catchphrase in a political speech or a straight play would break it up, and until they took Tommy's name out of the telephone directory fifty or sixty fans a night would ring him up to tell him that Fünf was after him, offering to do him now, and the like. Enabled by radio to share a pleasure of their people as never before, the royal family joined in. When Tommy explained after their command performance that *ITMA* was finishing its run but would return in the autumn, the Queen smilingly replied with the phrase of Ali Oop: 'I see. You go—you come back.'"

When the Show Could Not Go On

Of course, there is nothing inherently funny, witty, or even intelligible in the catchphrases from *ITMA* without the shared experience of the show, which in this respect could best be compared with examples a generation later: "And now for something completely different," from *Monty Python*, or "Sock it to me" and "Wery interrezting" from *Laugh-In*. The pervasive impact of a popular show such as *ITMA* is most remarkable in the theatre, which has always appealed to a more highbrow audience.

"Writers had to be careful," reminisced Ted Kavanagh, the New Zealander who conceived the show, "not to use parallel phrases, and worse than that was to give a cue which could expect an *ITMA* answer. No one can blame Shakespeare for taking risks of this sort, but it is a fact that *Macbeth* had to be changed because of an ITMA reference. An actor voiced the Bard's line, 'What is a traitor?' in just that tone that [the character] Naieve used. There was a burst of laughter, and next night the line was altered. Chekhov's *Uncle Vanya* got similar treatment when presented at the Westminister Theater in 1944. In one tremendously dramatic scene, with the audience at concert pitch to know what was happening, a shot rang out offstage. An actor rushed on, flourishing a still smoking revolver. 'Missed him,' he shouted, and the dramatic buildup was ruined as gusts of laughter swept the auditorium.

"Then there was the provincial tryout of Eric Linklater's play, *Crisis in Heaven*, at the Opera House, Manchester. Barry Morse played the part of Pushkin, the Russian poet, and in one of his long and rather pompous speeches he finished with the words: 'I go.' Just as if they had been rehearsed for the part, the audience replied with one voice: 'I come back.'"

A Pair of Aces

Catchphrases, dialect, and linguistic lapses all helped to fix a radio character in the public's mind. Jane Ace, the wife of Goodman Ace off and on the air, became celebrated for some of her malaprops:

- I got up at the crank of dawn;
- I was down on the lower East Side and saw those old testament houses;
- Be it ever so hovel, there's no place like home;
- The food in that restaurant was abdominal;
- Every picture I see of Abraham Lincoln makes him look so thin and emancipated . . .

Jane Ace would get so mixed up that sometimes at home she would have to ask her husband: "The fly in the oatmeal—or is it ointment?" Author Mark Singer, who happens to be Goodman Ace's grand-nephew, wrote a profile of him in *The New Yorker* in which he demonstrated how easily one could get lost in a linguistic forest around his great-uncle:

"The careful conversationalist might try to mix it up with him in a baffle of wits. In quest of this pinochle of success, I have often wrecked my brain for a clowning achievement, but Ace's chickens always come home to roast. From time to time, Ace will, in a jerksome way, monotonize the conversation with witticisms too humorous to mention. It's high noon someone beat him at his own game, but I have never done it; cross my eyes and hope to die, he always wins thumbs down."

Unto Us a Child Is Born

On January 20, 1953, 29 million Americans watched on television Dwight D. Eisenhower's inauguration as their thirty-fourth president. The night before about 44 million viewers were glued to their sets to watch the fifty-sixth episode of the situation comedy *I Love Lucy*, on which Lucy and Ricky Ricardo had a baby. The former was a routine event, one that happened every four years. Lucille Ball's confinement was a unique if controversial phenomenon that gripped the nation, and indeed, the media around the world.

The subjects of pregnancy and childbirth were considered by sponsors and network executives riskier in those days than abortion or child molesting would be today. Because *I Love Lucy* was already the top comedy show on television, CBS reluctantly allowed Lucy/Lucille to get increasingly larger, though the network exercised poetic licence by restricting the entire pregnancy to seven weeks. The word "pregnant" was found objectionable, and so the more tasteful "expectant" was used a few times in the dialogue. The first episode broaching the subject actually was entitled "Lucy Is Enceinte." (This might have worried American housewives who did not know that French expression, and looked up its original meaning: "enclosed.")

CBS executives were so worried about the moral damage they were about to wreak upon the public that they arranged a dress rehearsal for representatives of some of the major faiths—a monsignor, a rabbi, and a Presbyterian minister.

"Was there anything objectionable?" one of the network people asked anxiously afterward. In this the three clerics were found to be unanimous. "What could there be objectionable," they asked, "about having a baby?"

The press was flooded with letters, most of them in favor of Lucy's pregnancy. When one reader complained, another attacked her: "Has she heard of the story of Mary, Joseph and the Christ child?"

The Perfect Script

Behind the scenes at *I Love Lucy* arrangements for the birth of Ricky, Jr., were more complicated than just getting an unusual star to point the way to Bethlehem, which special effects could have handled rather easily. For one thing, two births were involved: Lucille Ball was really pregnant, and by what could only be considered supernatural coincidence, she would give birth to Desi Arnaz, Jr., on the same day as her on-screen character delivered Ricky Ricardo, Jr.

The episode "Lucy Goes to the Hospital" was filmed in the middle of November, when Ball went on a four-month hiatus. In those primitive days, neither the sex of the baby nor its precise birthdate could be predicted. The parents and producers decided that in the script the baby would have to be a boy, but the public relations people prepared different explanations for either sex. In fact, the release of information to the press about both the real and the fictional pregnancy was planned like a military campaign, designed to whip up maximum frenzy in the public. They learned on December 8th that Lucy was "enceinte," and only on January 14th that she would give birth five days hence on the 19th. "An electric current seemed to race through newspaper offices all over the country," writes Bart Andrews in his book about the show. "Perhaps Lucille Ball's real baby would make its appearance on the same day. Reporters everywhere picked up the telephone; hundreds of calls began flooding the Arnaz home, office, studio. . . . Newspapers published hourly bulletins and ran pools betting on the baby's sex."

When on the morning of January 19, Lucille gave birth on schedule, and to a son, Desi Arnaz exulted in the waiting room:

"That's Lucy for you. Always does her best to cooperate. Now we have everything!" Then he called the head writer of the Lucy show, and informed him:

"Lucy followed your script! Ain't she something?"

"Terrific!" Jess Oppenheimer responded. "That makes me the greatest writer in the world. Tell Lucy she can take the rest of the day off!"

Stranger Than Fiction

Almost forty years later, pregnancy on American television would still trigger a major debate, though of a different kind. Vice President Dan Quayle served up a feast for comics and political commentators when, in May 1992, he attacked the television character Murphy Brown for having a child all by herself. Robin Thomas, the actor who had played the baby's father and had been written out of the show, faxed Dan Quayle's office that he fully agreed with the veep and would come back to Ms. Brown at the drop of a hat. *Newsweek* columnist Eleanor Clift sprang to the sitcom's defense and said on television that Candice Bergen, the actress playing Murphy Brown, was used to dealing with dummies—after all, she grew up with Charlie McCarthy—and therefore would have no problem dealing with Quayle. And *Entertainment Weekly* commissioned a poll which showed that more Americans would vote for Murphy Brown as a parent or president than for Dan Quayle.

Alibi

Macdonald Carey, best known for his years on television in *Days of Our Lives*, first acted in soaps on radio. As a busy actor, who appeared on a number of shows, he kept an appointment book, which one day came to be used as evidence in a court of law. As Carey tells it in his recent memoirs, a friend called Craig asked him to provide an alibi after he had been accused of stealing a fur coat. The felony supposed to have been perpetrated in another city and another state on a day when Carey could prove, with the help of his appointment book, that he had lunch with the accused.

The prosecuting attorney was also interested in the datebook, but in order to discredit the actor's testimony.

"If you're going to believe Mr. Carey," he held up the little black book to the jury, "here is his day: 10 o'clock—Young Widder Brown. 12 o'clock—Craig and Marge. And at 3 o'clock, John's Other Wife."

The judge asked Carey what his profession was, and when he learned that the appointments—except for lunch—were the titles for radio shows, he dismissed the case.

The TV Defense

The most persistent opposition to television has come from groups and individuals who believe that violence on the screen causes violent behavior in society. Dade County, Florida, was the scene of the first trial in 1977 where the defense tried to use television-watching as the cause for a fifteen-year-old boy, Ronney Zamora, shooting an elderly woman next door.

Psychiatrist Michael Gilbert, an expert witness for the defense, argued that young Zamora pulled the trigger as a conditioned response, because of his addiction to violent shows such as *Kojak*. The bald New York detective, played by Telly Savalas, had become the young boy's hero, who even wanted his stepfather to shave his head.

In the end, the court rejected the defense's novel line of reasoning, causing attorney Elliot Rubin to complain:

"Your Honor, involuntary television intoxication is a new defense, but so was insanity at one time a new defense."

Though the state won its case, *Florida v. Zamora* proved to be a landmark trial as the first to be televised throughout the state. There was a great deal of controversy at the time, but it opened the door to such spectaculars as the William Kennedy Smith rape trial and Court TV, a cable channel which shows cases 24 hours a day. At first, the novelty did seem to interfere with due process, as when the jury in the Zamora case petitioned the judge if they could watch themselves on television during the trial with the sound turned down.

"They said they just want to see what they look like on TV," said Judge Baker, after turning down the request. And at one point, when both defense and prosecution attorneys appeared to ignore a ruling he had made, the judge complained:

"I only have a small role in this production."

Who Really Shot J.R.?

The television soap opera *Dallas* scored a publicity and ratings bonanza when the 1979–80 season ended with a cliffhanger: the principal villain, oilman J. R. Ewing, is shot by an unseen assassin. For the next six months, the world was obsessed by the one question: "Who shot J.R.?"

Scotland Yard came forward with its own theories, and British fans placed calls to Parkland Hospital in Dallas, to find out how J.R. was recuperating.

America, in the middle of the Iranian hostage crisis and a presidential election campaign, seemed to be obsessed with fiction instead. Jimmy Carter, on one of his few campaign swings, declared at a fundraiser: "I came to Dallas to find out confidentially who shot J.R. . . . If any of you could let me know that, I could finance my whole campaign this fall." Meanwhile Republicans were passing out buttons that claimed to solve the mystery: "A Democrat shot J.R."

At one point that summer Larry Hagman, who played J.R., announced that he would run for president, promising to raise the chief executive's salary to $5 million a year to get the right sort of greedy person to run for the office. J.R. had other programs, such as GI benefits for Veterans of Future Wars: "My idea is that government should pay for your education, set you up in business, give out all the GI benefits before there's war."

The real victims of the attempt on J.R.'s life were the programs scheduled by the networks against *Dallas*. When NBC's *News Magazine with David Brinkley* sank without a trace in the same time-slot, humorist Art Buchwald theorized that J.R. might have been shot by David Brinkley.

Son of Peter Pan

Broadway star Mary Martin was asked by a reporter once what it was like to have Larry Hagman, a living legend, for a son.

"No, dear," said the Texan-born actress who is known to millions of Americans as Peter Pan, "I am the legend. *Dallas* is a bloody cult."

Where Credit Is Due

Having the proper title or credit denotes power, which means that these are almost as important as money in Hollywood. Phil Capice, one of the executive producers of the popular television series *Dallas*, thought that having his name appear last among the credits would leave the most lingering impression, so he insisted that each show should fade out on his name. It became a standing joke within the production, and each week, when the roughcut was being assembled, new ways were devised to show the producer's credit in an innovative, if not wholly respectful, manner.

One week, all the credits had rolled, and it seemed as if Capice's name had been left off. Then Larry Hagman's face appeared on the screen, and as he flashed his best "J. R." smile, the camera went in close to reveal letters written on each of those gleaming white teeth, spelling out: P-H-I-L C-A-P-I-C-E.

A Posthumous Forsyte

The first truly global success the BBC had with a dramatic series was the television adaptation of The Forsyte Saga. *Kenneth More, who played the sympathetic central character of Jolyon Forsyte, recalls in his memoirs,* More or Less, *a quiet Christmas visit he had hoped to pass in Madrid with Angela, his soon-to-be third wife, whom he calls Shrimp.*

Our flight was delayed, and we arrived at the hotel at about half past nine in the evening, and checked in at the desk. Because we were tired, we decided to have a light meal in the hotel instead of going out to a restaurant. As we passed through the main lounge to the lift, I saw a large TV screen had been set up, with a number of people watching it. About fifty men and women in armchairs, eyes on the screen. The women wore diamonds and furs, and many of the men had grey hair and seemed the type who pass in advertisements as men of distinction. They were a pretty wealthy audience and all absolutely absorbed by the action on the screen.

I looked at the set to see what was holding their interest so closely. There I was, as Jolyon, speaking the most superb Spanish. I was grasping the curtains in

Jolyon's fatal heart attack, and died in front of an audience of fifty Spaniards in a Spanish hotel. A woman in one of the front seats, a middle-aged, dark-haired, diamond ear-ringed and fox-furred lady, was practically having hysterics.

"Es muerte . . . ! Es muerte . . . !" she cried in anguish. A friend in the next seat to her was fanning her and held smelling salts under her nose, and all because I had died on the screen. It was an extraordinary sensation for me. I had not realized the impact of *The Forsyte Saga* in Spain, and remember, this was the second showing! Shrimp and I exchanged glances.

"This is too good to miss," I said. "I'm going to tap her on the shoulder to prove I am still alive."

"Go on," agreed Shrimp. "Give her a laugh. She won't believe it."

On screen my eyes were closed for the last scene, as I began to pick my way through the chairs.

"Es muerte . . . !" cried the woman again.

I tapped her on her shoulder gently and said, "No, madam. He's not. He's *here*."

Pandemonium! The woman screamed and leaped out of her chair. The smelling salts were spilled and the glass bottle smashed. Everyone shouted in amazement.

Until then I had not realized we were the toast of Spain. Now, I was not allowed to forget it. Magazines displayed our pictures on their covers. The hotel proprietor insisted that everything we wanted must be on the house, and during our stay, when we went out to dine in a restaurant, someone at another table would invariably recognize us and insist on paying our bills. Whenever we waited for a taxi at a taxi-stand, we were immediately passed up to the front of the queue. I would protest, "No, *please*." But a Spaniard would reply in perfect English, with a smile: "No. No. Señor, please. After all, dead men do not argue." The whole experience was extraordinary, like a royal progress.

On Christmas Day, we decided to attend Mass in the oldest and most famous church in Madrid. We deliberately arrived slightly late because I did not want to be noticed.

We crept up a side aisle to one of the small chapels, because Shrimp wanted to light a candle to St. Teresa. No one saw us. Everyone was looking at the priest. Then a woman from the main body of the church turned and her eyes met mine. I thought, Please, God, not here! St. Teresa, please. Not *here*.

The woman immediately nudged someone next to her, who also turned to look at us.

"Jolyon Forsyte. Jolyon Forsyte," she whispered. Within seconds, the nudging went around the whole congregation and a kind of sigh ascended to the roof of the cathedral: "Jolyon Forsyte . . . Jolyon Forsyte . . ."

The old priest was about to chant when he saw me and went, "Ahhhh!" The whole service stopped. People now began to call from the back of the cathedral: "Jolyon! Jolyon!"

I panicked. There could be no Mass for us that morning. Shrimp and I literally ran out of the cathedral, and we did not stop running until we reached the seclusion of our hotel.

Extra

Before his career as a writer for Jack Paar and Johnny Carson, or as a talk-show host, Dick Cavett got on television mainly in walk-on roles. Once he talked his way into an episode of *The Phil Silvers Show* as an extra. At an opportune moment, Cavett introduced himself to Phil Silvers, reminding him that many years before he had got him to autograph a photograph during the Omaha tour of *Top Banana*.

"And here I am on your show," Cavett marveled at the amazing change in his own fortunes, but Silvers responded nervously:

"What for? To give back the picture?"

Fan-tasy

Years after the television series *Batman* has dispersed into the ether, actor Adam West still encounters female fans. One middle-aged woman offered herself at a shopping mall, saying, "God, you were my fantasy for years."

"Was I any good?" West asked.

I Don't Like You Just the Way You Are

James Arness, at the time he was playing Marshall Matt Dillon on the popular Western series *Gunsmoke,* sat in his convertible on the ferry crossing from San Diego Bay to Coronado. A woman in the next car spotted him.

"Matt Dillon!" she shrieked with delight. But then taking a closer look, her voice trailed off: "Why, you look exactly the way you are!"

Reality Check

Bob Hope was in New Orleans to tape one of his specials when a woman spotted him outside Antoine's, the city's famed restaurant.

"Bob Hope!" she exclaimed. "Is it really you?"

"No," Hope replied, "I'm on tape."

Pro Fan

At least one television fan, Dorothy Miller, became famous by attending tapings of her favorite shows. When Steve Allen was host of *The Tonight Show*, Ms. Miller followed him to Hollywood, Hawaii, and wherever the show toured. After a while, she became such an integral part of the show that she was forced to turn pro and joined the Screen Actors Guild.

Two Kinds of Fans

Comic Garry Moore, at the height of his television fame, was sitting on a park bench where a little girl recognized him.

"Garry! Garry!" she squealed with delight. Moore smiled and said: "Hi, honey." The little girl then burst into tears, and bawled him out: "You don't even remember my name!"

Moore developed a friendly way of dealing with fans, either waving at them or slapping them on the back. One foggy night in New York, a man poked a sharp object into his ribs, while mumbling something. Moore automatically greeted him with a hearty "You betcha, buddy!" and walked on. After a few seconds it sank in, and he turned around just in time to glimpse his would-be mugger bolting in the opposite direction.

Functional Illiteracy

A female fan was waiting for the English comic Norman Vaughan outside the ATV studios in Elstree. She pressed an autograph book at him: "My youngest will be thrilled; he's Val Doonican mad."

"I'm not Val Doonican," Vaughan corrected her. The mother paused momentarily and then said:

"Oh, go ahead and sign it anyway, love. He can't read!"

Harry Secombe, another English comedian, got tired of getting illiterate requests for "potos" to be "singed" by him. Finally, he lit a match and carefully scorched one of his glossy photographs around the edges, and sent it back to a fan with this note:

"Here is the singed poto requested. Sorry about its condition but we had a fire in the TV studios and it got signed all round the edges."

Tinker, Tailor

Deems Taylor, a well-known and serious composer and music critic, who also did many music programs on radio, was a great fan of Jimmy Durante. One day he came backstage after a broadcast, and asked Garry Moore to introduce him to his idol.

"Jimmy," said Moore, "I'd like you to meet Deems Taylor."

Durante looked puzzled, but put out his hand with a big smile, and said: "Whose?"

Role Model

Linda Ellerbee was anchoring *NBC News Overnight* when she received a qualified fan letter from a little girl, which said:

"Dear Miss Ellerbee,

When I grow up I want to do exactly what you do. Please do it better."

Mutual Fans

Some stars, who may not like importunate adult fans, relate well to children. My friend and lawyer Michael Donaldson recalls a charity event that he was attending at the Beverly Hilton with one of his clients, the late Michael Landon.

We were headed down the long hall entryway, talking about something or other, when a young child who had obviously undergone some chemotherapy slapped him on the leg, waving a pencil and pad. The size of the pad and the gesture of the pencil constituted a silent but fervent request for an autograph. Michael signed and knelt down to talk to the boy while his left hand thrust pad and pencil into the hands of the mother for safekeeping. He looked the boy right in the eye and said:

"Do you know who I am?" The poor kid didn't know anything. Michael picked him up and started talking to him about what it's like to be a writer and urging him to do some writing. After we headed in to the dinner, at which Michael was the honoree, I said to him:

"You genuinely like kids, don't you?" With seven of his own at that time, he did not have to prove that to the world. But his genuine connection with youth caught me for the first time that night.

"Yes," he said, "because they don't know who I am."

Hot Tip

Satirist Ogden Nash, who had tersely remarked that "Video is hideo," appeared as a panelist on *Masquerade Party*. He immediately noticed the impact of his newfound celebrity on his wallet.

"When a New York cabbie knows you from TV and calls you by name, your tip status automatically jumps. I used to tip a quarter. Now I have to make it fifty cents. If you're a public personality you cannot, after all, be a public cheapskate."

Jack the Tipper

Partly to counteract the image of stinginess he cultivated on the air, Jack Benny was a generous tipper. Once, just after broadcasting a skit in which Benny gave a messenger boy fifty cents, and explained that it was because he had forgotten his glasses, the comic took a cab. He gave the driver such a large tip that he gave him a punchline:

"By golly, Mr. Benny, you sure do need your glasses!"

George Burns remembers leaving Earl Carroll's nightclub with Jack Benny, who gave the hat-check girl a dollar, a sizable tip forty years ago. The girl handed back the bill, saying:

"Please, Mr. Benny, let me keep my illusions."

Hard Luck

Despite his miserly reputation, Jack Benny radiated a kindly personality which would bring him many hard-luck letters. "If you don't send me a certified check for $50,000 by return mail," one farmer wrote him, "I can't buy four tractors on which I've got an option. And if I don't get the tractors, I'll lose my whole farm. I never miss your TV show, and you have a kindly face, so I know you'll send me that $50,000. To prove to you I'm honest, I'll be glad to send you a photograph of the four tractors."

Brief Encounter

When Ed Wynn was down and out of work in the fifties, a fan recognized him walking down Madison Avenue.

"Ed Wynn," she stopped him. "I thought you were dead."

"I am," the sad-faced comedian replied. "Welcome."

CHAPTER 2

Out of Thin Air

The Story That Almost Got Away

Having patented the grid tube, which led to the three-electrode vacuum tube, Dr. Lee De Forest is generally recognized as the "Father of Radio" in a field where paternity is notoriously difficult to establish. More typically, there is a dispute about who sang first over the airwaves. A singer called Vaughn de Leath, who became known as "the original radio girl," claimed that De Forest's assistants had invited her to his laboratory; just before singing "The Old Folks at Home" into a phonograph horn, she is supposed to have said: "Well, here goes something into nothing." Ben Gross, longtime broadcast columnist for the New York Daily News, *checked this story out with De Forest himself, who denied any knowledge of it. In his entertaining memoir,* I Looked & I Listened, *Gross gives the official version.*

It was Dr. De Forest who broadcast the first musical program in 1907, from a tiny laboratory on the top floor of the Parker Building on Fourth Avenue and 19th Street, New York. Earlier in the evening, he had dinner with an old friend, a Swedish concert singer, Madame Eugenia Farrar, and her companion, a young newspaperwoman.

"If you'll come up to my laboratory," the Doctor told them, "I'll show you a wireless set which transmits the human voice over the air. In fact, you can be the first one to sing into it."

An hour later, in the laboratory, she faced an old-fashioned phonograph horn and into it the Swedish artist sang "I Love You Truly." It was the first song ever broadcast. "Encore!" Dr. De Forest shouted, and the Madame sang "Just A-Wearyin' for You."

She had scarcely begun her first number when in the Navy Yard in Brooklyn a young wireless operator sprang from his chair. "I hear a woman singing!" he exclaimed to the lieutenant in the room.

The officer, fearful of the enlisted man's sanity, clamped on the earphones. "I'll be damned!" he said, and then immediately phoned the New York *Herald*. His call was transferred to a rewrite man, and the latter, convinced that the newspaper was the target of a practical joker, called back a few minutes later to check on the authenticity of the information.

The lieutenant assured him that both he and the sailor were of sound mind. The rewrite man dashed off a brief story which was buried on page 5 of next morning's *Herald*. One of the most significant developments of modern times was given the treatment usually accorded an item of no more importance than a street-corner brawl. As for the young newspaperwoman who had witnessed the history-making broadcast—she regarded it so lightly that she did not even trouble herself to write a story about it!

Who's on First?

At the turn of the century, Thomas Edison told a friend that he thought some time in the future there might be daily signals across the Atlantic without wires, but that he did not know when, and being preoccupied he did not think he would have time to do it himself.

When Guglielmo Marconi did the trick in 1901, Edison said:

"I would like to meet the young man who had the monumental audacity to attempt and succeed in jumping an electric wave across the Atlantic."

Recounting the story of the historic transatlantic signal, Marconi said: "I shall never forget Mr. Edison's laconic comment: 'If Marconi says it's true, it's true.' Nothing ever pleased me as these words."

The Man and the Legend

David Sarnoff, arriving from Minsk in New York in steerage at the age of nine, apprenticed himself to Marconi at fifteen and quickly climbed the ladder of the latter's American organization. A self-made man who became chairman of the

Radio Corporation of America at thirty-nine, Sarnoff was no inventor except when it came to his own legend. The most indestructible one is how he became famous at age twenty-one, when he picked up the first distress signals from the ship that was trying to rescue passengers from the *Titanic*, and how he manned the Marconi station on top of the Wanamaker building during a lonely vigil of 72 hours, while the cream of New York society swirled around the building anxiously waiting news of their relatives and friends.

This story, with various embellishments, was told by Sarnoff and others over the decades, though Carl Dreher, an RCA engineer, pointed out that the Wanamaker wireless station was closed at night, when the first wireless message was flashed. And the Marconi company had, in fact, closed down the Wanamaker station during those crucial days to prevent interference with its four coastal receivers.

With his intuitive grasp of business and publicity, Sarnoff was a visionary who could foresee uses for the new flood of inventions that their inventors never imagined. During World War I, when radio-telephony was employed mainly by ships at sea, and any commercial development was blocked by the U.S. Navy's monopoly, Sarnoff wrote a famous memo developing a plan "which would make radio a household utility in the same sense as the piano or phonograph. The idea is to bring music into the home by wireless. . . . The receiver can be designed in the form of a simple 'radio music box' and arranged for several different wave lengths, which should be changeable with the throwing of a single switch or pressing a single button. . . . The same principle can be extended to numerous other fields, as for example receiving lectures at home, which can be made perfectly audible; also events of national importance can be simultaneously announced and received. Baseball scores can be transmitted in the air."

In effect, Sarnoff had visualized commercial broadcasting. Within four to six years everything he had predicted would actually come to pass, and the Radio Corporation of America went on to realize enormous revenues from both the invention and its long-held subsidiary, the NBC network.

After *Your Show of Shows* became a great money-spinner for NBC, Sid Caesar was invited to dinner in the private office suite of General David Sarnoff at RCA headquarters. At the end of the meal, Sarnoff said:

"Do you know that you make ten times more money than my top scientist down at our research laboratories in New Jersey?"

"Well, General," the thirty-two-year-old comic replied, "you have a great tank, the fastest, with the best armor and the most accurate gun, but if you don't have a driver, the tank doesn't go anywhere."

"I can always get a driver," Sarnoff insisted.

"Ah, but can he shoot, too?" asked Caesar.

Acrophilia

Edwin Howard Armstrong was a lone tragic figure among the early pioneers of radio. He found a way to send clear signals around the world and held a large number of patents, including one for wide-band frequency modulation, better known as FM. Armstrong licensed several of his inventions for great sums which financed further research, but most of his life was taken up by defending his patents against Lee De Forest and RCA, where he had worked for many years with David Sarnoff. He became convinced that Sarnoff and RCA were trying to break him financially and psychologically by endlessly prolonging the patent litigation, and the inventor committed suicide in January 1954.

Armstrong chose death by jumping out of his New York apartment, thirteen stories above the East River. It was characteristic of a man who had always loved heights. He had built his first antenna in 1910, when he was only nineteen, a three hundred foot wooden structure in Yonkers, from which he could pick up telegraph signals from as far away as Key West, Florida, and Newfoundland. In the most famous photographs of Howard Armstrong, the ebullient inventor is shown balancing on one foot atop a globe precariously perched on the RCA tower, four hundred feet above West 42nd Street in Manhattan. After he came down, one of the engineers asked him:

"Why do you do these damnfool things?"

"Because the spirit moves me," said Armstrong.

Atmospherics

Sometimes in the early days, broadcasting would be done from a tent, pitched on the station's roof, to take advantage of the best possible atmospherics.

A distinctive feature of KDKA's evening program was the whistle of a freight train which always sounded around 8:30, and became a regular sound effect for every show.

Harold W. Arlin, an engineer and announcer with KDKA, one of the first commercial radio stations in East Pittsburgh, introduced one night a singer, whose wide-open mouth attracted a flying insect. As the tenor lost control of his chords and polite vocabulary, the engineer pulled the switch.

Static

The big topic of conversation in the early days of radio concerned not the programs so much as the reception. Like the Eskimos' proverbial thesaurus of synonyms for snow, frustrated owners of radios came up with a slew of slang expressions to describe static ranging from the technical to the onomatopoeic.

"Wow-wow" characterized the howling noise caused by radiation from the antenna of a nearby radio; "crossfire" or "cross-talk" referred to sounds picked up from other broadcasts; "mush," according to the *Thesaurus of American Slang*, was "a frying sound"; "clinkers," "scratchers," "burp," and "bloop" were more or less what they imply. Interference, depending on what image the sound might conjure up, was also described as "termites," "whiskers," "birdies," or "canaries."

The Great Houdini got himself into more trouble than he bargained for when he went out and bought the most expensive radio set he could find. After he failed to receive anything but static, he took his purchase back and, in a fit of anger, smashed a number of other radios in the store. Hearing that a warrant had been issued for the magician's arrest, Alfred H. McCosker, a press agent who later became the head of the Mutual Broadcasting System, invited Houdini to tell his side of the story from the studios of WOR. The punchline of this publicity stunt occurred when the police arrested Houdini in the middle of his broadcast.

Complaint Department

In the 1930s static was still plaguing the Canadian airwaves to such an extent that listeners would make late-night telephone calls to the home of Hector Charlesworth, chairman of the Canadian Radio Broadcasting Commission, which later became the CBC. One night, after there had been a number of such calls, the chairman's Irish-born wife, Kathleen, listened to a man insisting that her husband do something about the static on his set, and then gave him some static of her own:

"Oh, I don't think it's Mr. Charlesworth you want. That's a matter you'll have to take up with the Almighty!"

On the Same Wavelength

In the twenties, before traffic on the air waves was controlled, programs from one channel would often seep into another, much to the listener's amazement. In a famous story evangelist Billy Sunday was thundering a sermon on one of the New York stations: "Brothers and sisters, I ask you what did Jonah say to the whale?" Then, the announcer from another studio cut in with the reply: "Take Carter's Little Liver Pills!"

Another celebrated evangelist on the other coast, Aimee Semple McPherson, delivering her sermons from her vast temple in central Los Angeles, deliberately kept changing the wavelength of her station, in order to attract new audiences. After many complaints by the rightful owners of those bands, the Department of

Commerce closed down her facilities. An enraged Miss McPherson fired off a telegram directly to Herbert Hoover, then secretary of commerce:

PLEASE ORDER YOUR MINIONS OF SATAN TO LEAVE MY STATION ALONE + YOU CANNOT EXPECT THE ALMIGHTY TO ABIDE BY YOUR WAVE LENGTH NONSENSE + WHEN I OFFER MY PRAYERS TO HIM I MUST FIT INTO HIS WAVE RECEPTION + OPEN THIS STATION AT ONCE.

What Big Teeth You Have

One of the most colorful characters among pioneer broadcasters was a large-boned, large-hearted, and large-mouthed Swede by the name of Nils Thor Granlund. Variously known as NTG, Granny, or just "Mr. Show Business," Granlund claimed friendship with everyone on Broadway and in Hollywood. As publicist for Loew's Theatres, he was entrusted by his boss, Marcus Loew, to launch and run a radio station, WHN, which later became WMGM, a publicity outlet for Metro-Goldwyn-Mayer.

Reception from WHN was erratic. The signal was rarely heard north of New York City, whereas to the east it was exceptionally strong out at sea. Of course, there were few listeners in the Atlantic Ocean, but those few certainly listened, as Granlund found out.

Along with most radio stations in the beginning, WHN employed mainly amateur talent, and that is how Granny boasted to have discovered and promoted everyone from Joan Crawford to Ethel Merman. He featured amateur singers and all kinds of performers and contests, and sometimes, when somebody did not show, Granny himself would read some of his favorite poetry by Poe, Kipling, and Robert Service. This was during Prohibition, and after a while Granlund noticed several suspicious characters, whom he recognized as henchmen of the notorious bootlegger Larry Fay. They would sit around the studio and ask for specific poems to be recited over the air. One night, Granlund ran into one of these characters at a nightclub.

"Been meanin' to tell ya, kid," the man said, "how much we like your poetry. Me and my crew go for that stuff."

"How come?" NTG asked incredulously.

"We got to like it," the racketeer replied. "That's how we listen for Fay's signals."

Certain poems requested by Larry Fay's "boys" became the prearranged signals for the rumrunners from Canada and Bermuda as they entered the three-mile limit at the dead of night and began to dodge the Coast Guard with their contraband. "I never heard anyone recite *The Shooting of Dan McGrew* or *The Raven* after that," Granny wrote in his book *Blondes, Brunettes and Bullets,* "without remembering that WHN's signal was heard more than three miles out to sea."

Anyone Out There?

With radio as a mass medium, it is difficult to imagine that in the early days hardly anybody was tuned in. And how would one find out? Ray Poindexter recalls how a fifteen-year-old Arkansas boy, Jimmie Barry, signed on at 5:00 P.M. each day.

"This is WGAR, the Southwest American station, located at Fort Smith, Arkansas, the playground of America. We have just come on the air, and we wonder if anyone is listening." He would call the names of various set owners and ask whether they were home yet. His telephone would ring, and someone would say that the program was coming in fine. When a person called and said that he was having trouble with his receiving set, Barry would announce that fact on the air and promptly lose most of his listeners. They would hurry to the home of the owner of the defective set to help with the repairs.

Premiere

Once radio broadcasting began in America, it spread rapidly. Publishing magnate Robert W. Bingham asked Credo Fitch Harris, working at his newspapers in Louisville, Kentucky, to launch WHAS in 1922. In his aptly named Microphone Memoirs, *Harris described the excitement of those early days.*

As our opening night approached, citizens almost raided electrical stores to buy crystal sets and earphones. Tube receivers had scarcely come into the broadcasting picture. A scattered few were built by budding young engineers (without loud speakers, of course), yet they spread out over so much room—or rather so many rooms—that few homes were physically able to house them. Crystal sets were fairly good while they worked. On going dead, the frantic fan would wiggle his wire whisker to another part of the crystal, or another, and still another. Then he might dash to the medicine chest and give it a dab of rubbing alcohol. If that failed he might put it in the oven for a ten minute baking. Meanwhile the concert was probably over. Those were good recipes in their day and generation, and during our first year of broadcasting we must have repeated them by telephone to a thousand anxious inquirers.

Carefully I had gone over my talent list and picked out the choicest material for our first big night, announced to open Tuesday, July 18, 1922. A cinema concern moved in to take a thousand feet of film. Some of it they shot in the afternoon, keeping a reserve for the first studio program. That reel, by the way, was later shown in almost every theatre of size throughout Kentucky and southern Indiana.

By half past six o'clock newspaper photographers, executives, departmental heads, reporters and a few especially invited others had arrived, and were very much in the way, standing around with mouths more or less agape while cautiously refraining from coming into contact with any metal surface. Will I ever forget it!

We were to open at seven-thirty o'clock. Yet by seven each singer and instrumentalist was placed, as well as eminent citizens who, according to pre-arrangement, were to be introduced to our great unseen audience. Then the movie men began to grind. Their klieg lights added to the dazzling ensemble, and to the heat. By seven-twenty the studio was closed. There we waited as the clock ticked off minute after minute. Mercury in the thermometer was about the only thing that moved. Then, one by one handkerchiefs appeared, but I frowned them back into their pockets. Handkerchiefs might make a noise.

Two minutes more to go! At the end of those torturous one hundred and twenty seconds a red signal light would flash on the studio wall, and we would be—on the air! I explained this quickly in a hoarse whisper, and once more warned the room to silence. No cough! No sneeze! My heart was pounding. Our star soprano was breathing painfully. I could see the contralto's pulse beating in her throat. All nerves were tuned to concert pitch. Suddenly the red light glowed! Some gave a little gasp. I, also, wanted to gasp, but swallowed it and exclaimed in my best manly voice: "This is WHAS, the radio telephone broadcasting station of the *Courier-Journal* and the *Louisville Times*, in Louisville, Kentucky!"

It was the first cry of our infant broadcaster. A rather long cry, but a lusty one.

I had rehearsed it at home until the family was almost crazy, and had further prepared a brief explanation of what our adventure hoped to accomplish. Those who lived more than a thousand miles away were asked to wire us, collect, if they heard us or not—a somewhat ambiguous Irish twist which got by without comment. But afterwards a destitute creature accused me of having said to send those telegrams "prepaid" instead of "collect." Maybe so.

Now came the moment for introductions of executives and a few broad-browed notables who were waiting in line, pale and perspiring. They looked frightened and forlorn. I recall thinking that they looked very much as I felt.

It was expected, as a matter of course, that when their names were called each would step forward and at least say "Good evening," or "How-do-you-do." But microphone-itis is a fearful disease, and as one after the other was presented he made only a low, courtly bow in the most—or nearly most—approved drawing-room manner, with never a word coming from his frozen lips.

Those silences were awful. They fairly thundered, seeming to shatter the calm air with ear-splitting roars. . . .

I did not sleep much that night.

Going in town next day I passed a church. On the bulletin-board out front was the subject of the pastor's following Sabbath sermon: GOD IS ALWAYS BROADCASTING.

Tinkerer

The early history of television had its share of eccentric tinkerers. One of them was John Logie Baird, a Scot who is often credited with the invention of television in the mid-1920s, though other claims go back to the late 19th century. Baird had pursued several entrepreneurial ideas before he turned his attention to television; artificial diamonds, undersocks, and Baird's Speedy Cleaner all preceded Baird's "televisor."

Interestingly, John Baird possessed little mechanical talent for a "garret inventor," as he is often called. He realized it himself, because he advertised in June 1923 in *The Times* of London: "Seeing by wireless. Inventor of apparatus wishes to hear from someone who will assist (not financially) in making working model."

Victor Mills, a wireless enthusiast, who met Baird that year, said the latter's workshop was filled with junk. It was Mills who helped Baird to obtain the first television picture, by putting his hand in front of the apparatus to check the light. "It's here! It's here!" he heard Baird yell out as the Scotsman saw an image of the hand appear.

On February 8, 1928, John L. Baird, who was more of a showman than a scientist, tried to emulate Marconi's feat of sending a television signal across the Atlantic. He transmitted the images of three people, including himself. One of the other portraits was that of Bill Fox, a newspaperman who had missed one of the great news stories, despite the fact that he was in it. After the images had been transmitted from London, Fox went home without news that they had actually been received in New York. The following morning he went to Baird's house, which was under a state of siege. When Fox finally called his office at the Press Association, his boss bawled him out for missing the biggest scoop in his life:

"Everybody wants to know whose face was seen on the other side!"

Presumably that face belonged to Bill Fox before any egg had settled on it.

In Hoc Signo Vinces

A more typical American tinkerer was Philo T. Farnsworth. An untutored twenty-one-year-old from an Idaho farm, he built an electronic television in

1927 which he demonstrated in Los Angeles. The young inventor already had a group of investors backing him. During a demonstration of the machine, a bank vice president kept asking: "When are we going to see some dollars from this thing?" Farnsworth said nothing but made a dollar sign appear on the screen.

Eye Am the Camera

Vladimir Kosma Zworykin, a Russian-born research scientist with Westinghouse, perfected the television "eye." Zworykin named it the "iconoscope." He first demonstrated it to Westinghouse officials in 1924, and publicly at a meeting of the Institute of Radio Engineers in Rochester, New York, on November 18, 1929. It had no motor or moving parts; scanning was done electronically.

Zworykin's progress might have been slowed by the attitude of his employers. After the first demonstration at Westinghouse, his manager asked a few questions, such as: "How long did you work on the system?" Then Zworykin saw him speak to the director of the laboratory. Later he found out that the manager had said: "Put this guy to work on something more useful."

Zworykin was the archetypal scientist, completely absorbed by his work. During World War II, while sharing a ride with fellow researchers, the car had a flat tire, which meant they would arrive at the laboratory late. Zworykin impatiently looked at his watch again and again. Finally when the hands were at 8:30, the usual time to commence work, Zworykin, resigning himself to his predicament, slipped his watch back into his pocket, tilted his head back on the seat, closed his eyes and quietly said:

"Well, gentlemen, let's get to work."

At one time a reporter asked Vladimir Zworykin:

"What do you dream about—electrons and all sorts of new wonders?"

"I sleep soundly," Zworykin replied with a broad smile.

Interviewed at the age of eighty-four, Zworykin admitted that the part he liked most about television, the machine he had midwived, was the off switch.

The Jury Is Out

Ernst F. W. Alexanderson developed novel mechanical scanners in the twenties. His projector reflected a cluster of seven lights on the screen and when associated mirrors on a drum revolved, the spots of light gyrated and whirled to cover the entire screen with light that "painted" the picture. He described it as "a multiple lightbrush system" and demonstrated it on December 15, 1926, at St. Louis. In

Proctor's Theatre at Schenectady, New York, on May 22, 1930, he projected seven-foot television pictures on a screen, flashed from his laboratory by radio. He used a perforated scanning disc and high-frequency neon lamps.

Alexanderson was asked at the time when television might be ready for the home. Never a man to publicly prophesy or speculate, he shrugged:

"Well, I don't know. What we wonder is if the public really wants television."

Regret

Edouard Branly was the inventor of the coherer, an essential element in broadcast technology. Irked by the use of radio for purposes of war propaganda, Branly was quoted from Paris on his 95th birthday as saying:

"It bothers me to think that I had something to do with inventing it."

Color Wars

Around 1950 there was a great battle between RCA and CBS over the introduction of color television. Peter Goldmark had invented for CBS a system that worked better than what RCA had under development at the time. General Sarnoff was anxious to sell many more millions of RCA black-and-white sets before opening a market for color. So he was determined to use any means to delay or blunt CBS's technological edge. He told newspapers that the rival system was just a Rube Goldberg device, like "a bicycle and ricket chopper hauling a small suitcase from the fireplace to the hall closet." Interior designers were suddenly reported at a loss how to incorporate the CBS color sets into the decor.

Newspapers carried dire warnings for viewers, and people in the television industry became anxious when they read that platinum blondes would suffer in color, not to mention having to buy new decor and costumes. Fred Allen, the great radio comic, who was having problems with the transition from radio to even black and white television, dismissed the hullabaloo about color.

"It won't bother me," he said. "I don't blush."

Eureka, Alas!

Peter Goldmark, a Hungarian-born scientific genius, started working for CBS in the 1930s. He invented the long-playing record, which replaced the old 78 rpm disc and made millions for CBS. Following that expensive debacle over losing the color-TV war to NBC, a CBS vice president came to a board meeting to announce the latest breakthrough from the fertile mind of Peter Goldmark. William Paley, founding chairman of CBS, reportedly turned pale and gasped:

"No, not another Goldmark invention. I can't afford it."

Eureka—Almost

Before the invention of TelePrompTer, newscasters faced the problem of having to look down at their script, instead of the camera. Don Hewitt, later famous as the creator of *60 Minutes*, was the producer of early CBS experiments with weekly television newscasts, featuring Douglas Edwards. After trying cue cards, which made Edwards shake his head sideways instead of bobbing up and down, Hewitt rushed in one day with a new idea.

"I've got it! I've got it," he yelled at Doug Edwards. "You'll have to learn Braille. Then you can finger your script while looking into the camera at the same time."

Why We Have Zoom

In the early days, television cameras had to be tracked forward and pulled back to get the same effect that zoom lenses produce today effortlessly. Eric Fawcett, a BBC producer for Light Entertainment programs, was supervising from an upstairs gallery the shooting of a can-can routine with the Windmill chorus girls, urging on his cameraman: "In, in! Go in!"

Vera Seton-Reid, a vision mixer for the program, recalls hearing a small voice say: "Mr. Fawcett, if you have that camera track in any more, I'll have you up for rape."

The Rewards of Scientific Inquiry

The Playboy Channel might have had its origins in the 1930s, but the discoverers of R-rated television behaved with British decorum that would have made Superman proud.

"Scientific progress took an unforeseen turn yesterday," the London *New Chronicle* reported in August 1933, "to the embarrassment of two BBC men who were experimenting with infrared-ray television. The engineers were 'looking in' at a row of dancing girls who were being televised when, to their astonishment, they noticed that only one of the girls appeared to be clothed."

The explanation of this natural phenomenon is that infrared rays penetrated past cotton, but were stopped by the silk worn by one of the girls.

Olga Vysotskaya, a personality on early Soviet television, which began experimental broadcasting from Leningrad in the summer of 1938, was shocked by the number of letters she received following a gymnastics class on the air. Viewers wanted to know why she insisted on doing her demonstration in the nude.

My Favorite Martian

White did not show on early experiments with television, so in the 1930s everything white had to be painted green to look normal. After three Soviet aviators
flew over the North Pole for the first time, they were to be interviewed by Olga
Vysotskaya. One of the airmen was a little late into the studio and she greeted him
already in her Martian makeup. When the man looked ready to throw up, she
tried to calm him down:

"You, too, will look like this in a minute."

"Not on your life," said the Soviet pilot.

CHAPTER 3

Live or Dead!

Beyond Control

One of the more embarrassing liabilities of live television is called "cross-talk," when instructions from the control booth reach the audience instead of the intended technical crew or performer. This happened often in the early days before tape, and the feedback sometimes was not just of the electronic kind.

Ken Whelan recalls in his memoirs, *How the Golden Age of Television Turned My Hair to Silver*, one incident that contributed to the title of his book. As associate director at CBS on an episode of *The Red Buttons Show*, his job consisted of telling each of the three cameras during the live filming where to go next and whether to use a tight, medium, or wide shot.

"Two to dining room! Three to the bathroom! One to bedroom number two!" and so it went, until Whelan saw from the control booth that camera number three was not moving toward the bathroom. "Three, let's go! Three, what the hell's the matter? Go to the bathroom! Go to the bathroom!"

Finally, the second cameraman informed Whelan through his headset that camera three had been cut off from communication and could not respond.

Somehow they got through to the end of the sketch, by which time all the phones at CBS were ringing. Even after learning that he had been the victim of the much-dreaded phenomenon of cross-talk, Whelan could not quite understand why one of the sponsors who called sounded so apoplectic. Finally, he remembered the repeated instructions he gave toward the bathroom, when one of the CBS corporate directors, somewhat drunk, phoned to congratulate him on a beautiful program:

"Young man, it's just possible that you have caused the greatest mass bowel movement since the invention of Ex-Lax."

The Lord Helps Those

Some years ago Channel 11 in Hollywood used to broadcast a live religious program called *Churches of the Golden West*. A crew with a mobile unit would show up each Sunday morning at a different church to televise the service. It was a relatively simple operation, with the director connected to his assistant, crew, and narrator by a closed-circuit sound system, which was also patched into the public address system inside the church, if any instructions needed to be given to the congregation before the program began.

One Sunday morning, as Richard Webb and Teet Carle relate in their book *The Laughs on Hollywood*, director Joseph Agnello had set up everything inside a Protestant church in San Bernardino, an hour from Los Angeles, and the program began with its introductory titles.

"Cue the minister," Agnello ordered his assistant over his headset. Nothing happened. He could see the minister standing near the altar still waiting for his cue. The director twice repeated the instruction to his A.D., again with no result. Meanwhile, unable to hear, the assistant director spoke, and his voice reverberated through the entire church by means of the public address system:

"When do I cue the fucking priest?"

The minister finally heard that cue and quickly began the service. The show had a Catholic priest who acted as a commentator and religious advisor. His gentle voice now came over the director's headset:

"Joe, he's not a priest!" he whispered. "The line should be: 'When do I cue the fucking minister!'"

The Golden Age

Toward the end of his life Andrew Allan, often called the father of radio drama in English Canada, received a phone call from a graduate student who was writing a dissertation on the Golden Age of Radio. He wanted to come and talk to Allan.

"What Golden Age of Radio?" asked the producer.

"The one I am told you were intimately connected with," said the young man pleasantly.

Allan dodged the interview but that night he could not go to sleep. He tried to recollect what was so golden about that bygone era. "I always notice," he observed, "that when people celebrate that Golden Age, they dig up programs to represent it that are not only not golden, they're not even base metal. To those of us who believe the public was worthy of something better than soap-opera, those were the programs we were bucking against.

"When anyone refers to those years, as 'the Golden Age of Radio,'" Andrew Allan concluded, "it might be well for him to reflect that no age is golden until it is past. While it is going on, it may be as much as vexation as anything else."

The Aluminum Age

Back in the fifties, during what is often referred to as the golden age of drama in American television, censorship exercised directly by the sponsor was both heavy-handed and often ludicrous. Rod Serling received every conceivable award for his script *Requiem for a Heavyweight* (1956), but not before he was forced to change an innocent-sounding line like "Got a match?" The sponsor happened to be Ronson, the people who make lighters, not matches.

In another one of Serling's shows, which showed the New York skyline through an office window on stage, the Ford Motor Company insisted that the familiar Chrysler building be painted out. A cigarette manufacturer substituted such loaded words as "lucky" or "American"—both rival brand names—with "fortunate" or "United States."

While such interference might have been trivial irritants, other objections managed to turn gold into lead. These corporate censors drove Rod Serling into science fiction and, ironically, became directly responsible for *The Twilight Zone*, which has ensured Serling's posthumous fame more securely than any of his serious dramatic pieces.

The turning point came, as writer Marc Zicree tells it, with a teleplay Rod Serling wrote in 1956 for *The U.S. Steel Hour. Noon on Doomsday*, as originally conceived, was about an old Jewish man whose murderer is acquitted by a small-town jury. Word leaked out about the plot before it was aired—live, of course—and a reporter asked Serling if he was basing the story on the famous Emmett Till case, in which the white murderers of a fourteen-year-old black boy were acquitted by an all-white Mississippi jury. "If the shoe fits," is all that the writer replied, but the media made such a fuss about it that U.S. Steel received three thousand letters threatening a boycott unless changes were made. Serling was summoned to make the changes, which eventually resulted in moving the venue to somewhere in New England, with the murdered victim no longer Jewish but an unnamed foreigner. Told about the boycott, Rod Serling protested with ineffectual sarcasm:

"I asked the agency men how the problem of boycott applied to the United States Steel company. Did this mean that from then on all construction from Tennessee on down would be done with aluminum?"

Mystery

A radio murder mystery drama required the sound of an axe crushing a human skull. The sound effects people tried a variety of methods, chopping cabbages, pumpkins, and lettuce, but the producer was still dissatisfied. Finally, as the axe sliced through a watermelon, the producer said:

"Ah, that's it. It sounds just right."

"How do you know?" asked one of the sound engineers.

How to Hear a Pin Dropping

The script for one of the *Blondie* shows called for a pin dropping on the floor. Everything was tried, including a pin, which the microphone could not pick up. Finally, producer Don Bernard suggested dropping a railway spike on the floor. "We'll cut down the volume in the control room." And that worked.

Noises Off

Peter Duncan, who later hosted the popular interview program In Town Tonight, *began his career in broadcasting with a bang.*

I truly started at the bottom in the BBC—banging coconut halves on the floor to make the noise of horses hooves at £2 a week. Nowadays, the majority of effects are recorded, and I helped to make many of these, including the celebrated water splash, which was "canned" by an astute recording engineer just as I fell out of the boat into the Thames at Windsor while we were recording effects for a broadcast of *Three Men in a Boat*.

But in the early 1930s, we effects operators, including Brian Michie, George Inns, and several others who have since made their names in the entertainment world, could often be seen around Broadcasting House carrying odd bits of tin and wood, and experimenting in sound. Sometimes, when there was great activity in the studio such as a shipwreck or a highway robbery, we just didn't have enough hands, and had to call in members of the cast to help. I shall never forget

Sir Ralph Richardson chasing Sir Laurence Olivier round the studio with a whip that I had harmlessly asked him to crack at the appropriate moment, in the days before either of them were knighted.

I was still dropping trays of china for "noises off" when I first met an unknown radio actor, Leo Genn. Leo preferred acting to practicing law, for which he had a degree—a degree which was later to lead him to conduct the investigation into the notorious Belsen camp, and act as assistant prosecutor at the subsequent trial. I was glad to be able to welcome him in *In Town Tonight* when he had been starring in the film *No Place for Jennifer*.

Making "noises off" was great fun, particularly in *Band Waggon*, which brought radio fame to that "big" little man, Arthur Askey, and Dickie Murdoch, who started off as his stooge and stayed to become himself one of radio's leading comedians.

Both of them have broadcast in *In Town Tonight*, and both reminded me of the occasion I nearly knocked them out, right in the middle of a broadcast.

Harry S. Pepper, who was producing *Band Waggon* with Gordon Crier, wanted a crash to end all crashes. I can't remember if Arthur and Dickie were to fall from a skyscraper or go through the roof of their flat; whatever it was, a phenomenal noise was essential.

A colleague and I spent a good couple of hours building up a stack of tables and chairs, odd bits of wood, crockery, tin cans, anything we could find that would be guaranteed to make a noise. We balanced the whole lot on one table leg, so that a casual push would bring it toppling down, and while he held a whistle in one hand and banged a drum with the other, I used my free hand to clash cymbals; the other clutched a script. A good time was, indeed, going to be had by all.

Came transmission. Arthur and Dickie were standing in front of the microphone facing the audience, their backs to my contraption. They led up to my cue, and I waited, script in hand, until, at the exact moment, I shot out a foot to dislodge the table leg and sent the lot crashing on to the floor. How I stopped myself yelling I don't know, for the whole pile swayed precariously for a second, and then tottered over the wrong way. Instead of going harmlessly down at the back of the stage, it veered towards Arthur's and Dickie's unsuspecting backs, as they stood before the microphone. Down it went, missing them by inches. . . . Arthur, imperturbable as ever, looked calmly down as a tin can rolled in front of his feet. "Any old iron?" he asked, conversationally, as the St. George's Hall audience enjoyed one of the biggest laughs they had ever had.

Praised Be de Lawd

The Theatre Guild produced on radio Marc Connelly's exuberant gospel version of the Bible, *Green Pastures*. One of the actors was so nervous that he fumbled and finally forgot his lines. Juano Hernandez who played the principal role of De Lawd rose to the occasion and rescued the struggling mortal.

"Son," he boomed, "you is nervous before me and I can understand that. But I is de Lord, and I knows what is on your mind."

Then he calmly recited the missing lines and saved the moment.

Desperate Measure

In the early television soap *One Man's Family*, Eva Marie Saint was supposed to be talking to a fellow passenger on an airplane. But her partner suddenly forgot his lines, looked around, and said, much to her astonishment:

"Excuse me—this is my stop," and he made a hasty exit from the plane.

Keep in a Dry Place

Memorizing lines presented an enormous problem for actors in the early days of live television. Unlike the theatre, a familiar place with prompters and adequate rehearsal periods, the studio belonged to the technical crew handling the cameras and lighting, in which performers were simply in the way.

In radio there was the script; in the movies, scenes were very brief, and, of course, there could always be another take. On a live broadcast, the show had to go on.

Fortunately, the equipment failed so often that Christine Hillcoat, a makeup artist with the BBC, remembers an elderly actor telling another elderly colleague not to worry. "If you dry, old boy, just carry on mouthing words. They'll think it's a technical fault and they've lost sound."

Raymond Francis, the star of a British series *No Hiding Place* on Associated Rediffusion, had a huge number of lines to memorize. He used the old stage trick of hiding cards containing various speeches all over the set, so he could incorporate them as part of his natural perambulations during the play.

Bob Service, a veteran cameraman, remembers a nasty practical joke during one episode. Noticing that Francis had hidden one of these pieces of paper in a desk drawer, some pranksters substituted it while the actor was being made up. During the live show he reached for the drawer, but the card now read: "You will dry now." Which he did.

Existentialism

Ed Wynn forgot his lines on one of his shows and could not see the cue card. "I must have something to say," the old vaudevillian improvised, "otherwise I wouldn't be standing here."

Transcendent Moment

Fred Graham, longtime legal correspondent for CBS News, remembers one of his early days on the network, when he almost dried up. Walter Cronkite had already introduced him, when Graham happened to glance at the small compact of powder that the makeup lady had left in his hand. "I realized for the first time that I had been appearing on television each night before twenty million people wearing a face powder called 'Gay Whisper.'"

Good to the Last Drop

In a commercial break during Red Buttons's variety show, the announcer tried to pour steaming Maxwell House coffee into a cup which split in half. The show was live and Red Buttons immediately stepped into the breach.

"And no other coffee can make that claim," he adlibbed.

You Ain't Seen Nothin' Yet

Al Jolson suffered from stage fright all his life. To overcome his nerves for his radio broadcasts, he would enter the studio through the audience and shake hands with people as he progressed down the aisle. Sometimes he carried a picture of his infant son, and he would show it to individual members of the audience, "This is my little baby, this is my little baby." Once, he had Groucho Marx as a guest who followed Jolson on this trip through the crowd. Except Groucho was showing a photograph of a beautiful, barely-clad woman, and he was telling the audience:

"This is really his little baby, this is really his little baby."

Pasta Envy

Lucille Ball was observing Dean Martin perform for a whole hour on live television without any rehearsal. "That son of a bitch," she signed enviously, "makes cooked spaghetti look so tense!"

Twilight of a God

Carroll Carroll, who produced several radio shows sponsored by clients of the J. Walter Thompson advertising agency, scored a coup when he got John Barrymore to perform in a radio drama in a program called Shell Chateau. *The Great Profile was in the final phase of the disease that killed him.*

This was at the time in the affairs of man, particularly Jack Barrymore, when he was playing a real life Caliban to Elaine Barrie's Ariel. Their road to romance was a rocky one and every bump was detailed in every daily. Their antics were considered by many to be shocking but the juice John was using didn't come from Con Ed.

The agent who booked Barrymore claimed to have made a deal with Elaine, who said she had Jack out at her parents' home, to see that he was delivered to NBC for all rehearsals and the broadcast on time and reasonably sober. The way it turned out was that Jack pulled a Houdini and was finally found (who knows how he got there) locked in a room in the Hotel New Yorker, wearing nothing but a bathrobe and demanding that the girl on the switchboard connect him with Bacchus. He was brought to rehearsal and taken under the wing of Herb Polesie, who knew if he didn't keep him in tow *Shell Chateau* would go on the air *ex*-Barrymore.

The scene John was to play was from *The Jest*, a big Broadway hit in which he had costarred with Lionel. An actor named Royal Beale was doing the other Barrymore role. Polesie, directing John during rehearsal, said, "I wish I could get Royal to play the scene the way Lionel did, you know, real shit-in-the-barrel."

John said, "I don't think you could get Lionel to do *that* anymore."

When the time came for Barrymore's appearance, Jolson made a suitable announcement as Beale, Sylvia Field, and Edward Emerson took their places at the cast mike. Barrymore's was downstage center. To get to it he had to make his way through a forest of chairs on each of which sat a musician.

Watching from my place in the control room at the back of the studio, I could see John from the moment he stepped onto the stage. Doctors had been attending him and the questions were, did he have enough in him to keep him "up" long enough for the part or did he have enough in him to lay him low and how would the medicine they'd given him work? I waited and finally I saw him start to make his entrance. Then I heard the sound of a foot tripping over a cable and a "distant" grunt that, at the high level of the control room, was definitely "God damn!" But John kept coming.

When he reached the microphone he took hold of the stand and just stood there for a moment swaying, his other hand trembling so badly I was sure he wouldn't be able to read the words on the script he was clutching. Norma, who was sitting in the first row, said the audience gave out a mass gasp at the way he looked. His color was pale green and perspiration literally poured from his body until by the time the twelve-minute spot was over the part of the stage on which he was standing was not merely damp. It was thoroughly soaked. Everyone in the audience was sure John was about to die. But to the radio listener his voice was deep, resonant, and beautiful.

Suddenly he started to sway more than he'd been swaying. He was holding onto the mike stand so the mike swayed with him. He began to breathe heavily. I couldn't watch it anymore. I lay on the floor of the control room to get a "broadcast effect" and spare myself the awful sight of a great actor falling apart.

Suddenly, as he was reading his lines, he began to throw in incoherent asides. Then his glasses fogged up and he couldn't see. "Where are we?" he muttered desperately to Sylvia. Eddie Emerson ran over and whispered a line in his ear and pointed to his place on the paper. Then he dropped two pages of the script. Sylvia dropped two of her pages and read right along with him. The twelve minutes ended in a tremendous outpouring of applause for a man 250 people had just seen return from the grave. Barrymore bowed and walked off the stage, God knows how, without falling down.

John Reber called from his farm near Reading, Pennsylvania, to say that Barrymore had just given the most sensational performance he'd ever heard on radio. We told him what had gone on and asked him if he wasn't bothered by the snorting, grunting, and heavy breathing. "It sounded like passion," said John.

True West

In the summer of 1958 NBC produced a ninety-minute television program about the history of Westerns. Directed by John Ford, it was broadcast live from Gene Autry's Melody Ranch in the San Gabriel Mountains of Los Angeles. On the day of the broadcast, John Wayne, one of the principal characters, turned up late for the final rehearsals, nursing a hangover. Ford, one of the few people whom Wayne worshiped, made the actor sweat through several re-enactments of the famous showdown in *High Noon*. Wayne meekly obeyed, but was clearly performing under par. So Ford drew aside Autry and said:

"Gene, about twenty minutes before we go on the air I want you to give Duke a good healthy slug of Bourbon. And halfway through the show give him another. But don't let him know that I know."

As showtime approached in the afternoon, Autry went to his ranch house and filled a large bottle of Coke half full of Bourbon, and offered it to Wayne, who was panting and parched.

"Here, Duke," said the Singing Cowboy, "have a swallow of this. I think it will help you."

"Get that mouthwash away from me," Wayne waved at the Coke bottle, "you want to get me sick?"

"Duke, I'm telling you," Autry persisted with extra emphasis. "this is just what you need. Try it. Take a sip."

"He looked at me a little suspiciously," Gene Autry recalls, "took the bottle, and tilted it to his lips. His eyes got as round as saucers. When he handed me the empty he said, 'Autry, you may have saved a man's life.'"

Years later, after John Wayne had made *True Grit*, he sent Autry a photograph of himself in the character of Rooster Cogburn, with a black patch over one eye. He had written on it:

"To Gene Autry. A lot of water has gone under the bridge. And whiskey, too."

How the Pros Do It

Dick Whittinghill, for many years the morning disc-jockey on Gene Autry's KMPC in Hollywood, worked briefly for KIEV in neighboring Glendale, where he had a program beginning at 1:10 in the afternoon. The first day Cal Cannon, who owned the station, came by and asked his new employee to lunch. The meal was so liquid that no food was consumed at all. At one o'clock Whittinghill staggered to his feet, remembering he had a broadcast in ten minutes, and said:

"Mr. Cannon, I don't think I can do this."

"You're a pro, aren't you," the boss taunted him.

"I don't think so," said the drunk broadcaster.

"In order to be a pro," Cannon instructed him, "you've got to go on the air, drinking."

The Curse of the Drinking Class

Drunken actors have provided the theatre with some of its best anecdotes. Nowadays the problem is controlled with retakes and editing. But in the days of live television there were some anxious moments, as the British producer Brian Tesler told Denis Norden about one guest panelist.

There was an early-fifties panel-game called *Guess My Story* where the regular panelists were Pat Kirkwood, Helen Cherry and Michael Pertwee, and in the fourth seat we'd always have a guest panelist.

One memorable Thursday the guest was Gerald Kersh, the novelist. He was fine at the rehearsal, but during the break before the actual show I was eating in the canteen and the make-up lady came in and said she'd just been in to make Gerald Kersh up and he was absolutely pie-eyed. This was an hour before we were due to go on the air live.

I got hold of Larry and Pauline Forrester, our researchers—he's now a screen-writer in Los Angeles—and sent them over to sober him up. They poured black coffee down him, walked him up and down, just like in the movies, and took him out for some fresh air, but it didn't help. I had him sitting between Helen Cherry and Pat Kirkwood and though they tried as hard as they could to find the answers in the thirty seconds of questioning each panelist was allowed—so that I could skip putting the camera on Kersh—it didn't always work. When the bell went and I had to go to him, he'd say something like, "Whosh shosh a person was it?" All I could do was move straight to the next panelist. In his fuddled way, he was obviously aware that all he was getting was about three seconds each time, but luckily he couldn't quite make out what was happening and we managed to get through to the end of the show.

That was where each panelist said a good-night to the audience, but as we cut to Gerald Kersh for his the sound went absolutely dead. I yelled, "We've lost sound—what's happened?" But by then we were off the air and the sound supervisor came in and said, "I'm awfully sorry, there wasn't time to check with you, but when the camera got to Gerald Kersh the way he leaned forward into his microphone I thought he was going to say something bad, so I switched his mike off."

I thanked him, went down to the studio floor, and when I got down there everyone was aghast, because they thought what Kersh had said had gone out. He'd leaned forward and said, "I just want to say I would have been much better in this programme if it hadn't been for my attack of worms."

Loose Canary

In one of the early demonstrations of television's global reach, Edward R. Murrow brought together in November 1959 Irish playwright Brendan Behan in Dublin, comic Jackie Gleason in Philadelphia, and the critic John Mason Brown in New York. The CBS program was called, appropriately enough, *Small World*. Behan, drunk as usual, was unruly, producer Fred Friendly recalled: "When the program began and the boom man swung his microphone out of camera range, Behan rose in hot pursuit; every time it moved away from him he shouted, to the amusement of millions of viewers, 'No, you don't . . . come back here,' chasing the mike as though it were a canary on the loose."

During a break, Behan simply left, and failed to return, which Murrow attributed to the usual "circumstances beyond our control." John Mason Brown commented, "It's not an act of God, it's an act"; but Jackie Gleason decided to tell the truth:

"It was an act of Guinness."

Poets at Work

Dylan Thomas was just nineteen when he made the first of many broadcasts on the BBC, reading fifteen minutes of his poetry. In a sense, he was the first modern poet whose reputation was made through the power of his voice on radio. At first Thomas's nerves drove him in the natural direction of the public house, and in his early career on radio there were some hairbreadth escapes. In his recent book, On the Air with Dylan Thomas, *my friend Ralph Maud cites another poet, Roy Campbell, who produced several of the young Welshman's broadcasts.*

I used to keep him on beer all day till he had done his night's work and then take him down to the Duty room where the charming Miss Backhouse or Miss Tofield would pour us both a treble whiskey as a reward for our labors. It was with Blake and Manley Hopkins that Dylan became almost Superman: but we had bad luck with Dryden. Dylan had got at the whiskey first and he started behaving like a prima donna. He insisted on having an announcer instead of beginning the program right away as we used to on the Third Program. There were only two minutes to go and I rushed back to the studio and found Dylan snoring in front of the mike with only twenty seconds left. He was slumped back in his chair, with an almost seraphic expression of blissful peace. I shook him awake and, to his horror and consternation, began announcing him in my South African accent, but trying to talk like an English announcer, with my tonsils, in an "Oxford accent." Dylan nearly jumped out of his skin with fright and sobered when he got the green light, though he did bungle the title as *Ode on Shaint Sheshilia's Day*; but after that, his voice cleared up, and I began to breathe again.

When he had finished reading the *Ode*, I got another fright: he began to beckon me wildly with his arms and point to the page before him. I got the engineer to switch off the mike and slipped into the studio again. Dylan had forgotten how to pronounce "Religio Laici." I told him and slipped out. He had about three shots at it, bungled it, gave it up; and then went on reading. The next day I was hauled up in front of George Barnes, but he was a good boss and had a sense of humor. I promised to keep an eye on Dylan: Dylan promised to keep an eye on himself—and he kept his word.

The Time of His Life

One of the most prized qualities on live television was the ability to proceed in the face of unexpected disaster. An extreme case was Jackie Gleason, who did a whole hour's variety show by himself in front of a curtain that refused to go up.

In one of his dramatic roles, in a *Playhouse 90* production of William Saroyan's *The Time of Your Life*, the Great One discovered, probably to his delight, that the stage champagne had been substituted for something much stronger.

"I was half stiff!" Gleason told critic Tom Shales: "We all were. A young kid comes in and sings some Irish song and we're all supposed to cry, and we really did cry. We blubbered like babies."

Nobody noticed the reason, and afterward Saroyan called the production the best he had seen.

The Rest Was Silence

Broadcasting, even more than nature, abhors a vacuum: every second has to be filled with either sound or image. Martin Esslin, former head of BBC Radio Drama, recalls an incident when the air went dead.

During World War II, the BBC drama section was evacuated from London, and faced the problem of finding actors to fill the shoes of both those who were away fighting and those who refused to follow the Corporation to Bristol. Eventually, a repertory company of some 30 to 40 actors was established. Because the war kept people at home, especially in the evenings, radio drama became enormously popular, and as a result several of these previously unknown actors became household words. Some of them got swollen heads, and they were distinctly less popular with BBC staff members than with the public.

One such actor was Norman Shelley, who became known for his self-important and officious manner. After the war he was playing some role in a live production of *Hamlet*, which the director had not rehearsed or timed properly. Val Gielgud was the head of Radio Drama at the time, and he had warned that at 10 o'clock the network would have to switch to a live concert from Glasgow. However, when he saw that the play would run over by a considerable margin, he consulted with Glasgow and decided to send the Scottish Symphony Orchestra home. Norman Shelley knew about the original warning, but not about Gielgud's reprieve. So, as the hour of ten approached, and Hamlet was still chatting with the gravediggers, Shelley suddenly stepped in front of the microphone, and in his most solemn, pompous voice, intoned:

"And now the time hath come for us to bid adieu to Elsinore . . ."

Confusion reigned in the studio, and a long silence in the ether.

Nearly Strangled

John Snagge served with the BBC for almost six decades, from 1924 to 1981. "He became an announcer at Savoy Hill in 1928 in the days when they all had to put on dinner jackets to read the news," his longtime colleague Brian Johnston recalled:

"They also used to have a pianola in the studio, which the announcer was responsible for playing to fill in occasional interludes. One day John was off-duty in the rest room when he heard strangled noises coming from the studio. He rushed in to find that the tie of one of his colleagues had caught in the pianola roll, and he was gradually being pulled into the pianola and strangled to death.

"Later on Ernest Lush took over from the pianola and used to play the piano to fill in. Stuart Hibberd—the announcer known as the 'Golden Voice'—is reputed to have said, 'There will now be an interlush by Ernest Lude.'"

People Who Need No Introduction

One of the early American radio announcers, Harry Von Zell, made the following introduction: "Ladies and gentlemen, the President of the United States—Hoobert Heever."

Then, according to some versions of the legend, the announcer corrected himself with "Hervie Hoober."

On another occasion, Von Zell said: "My dear friends, the Duck and Doochess of Windsor."

Art Whiteside once welcomed visiting royalty to his microphone: "Today it is our extreme pleasure to introduce the Brown Quince of Norway."

CBS sent correspondent Robert Trout to cover for American audiences the first broadcast made by Pius XII. He introduced the Pontiff as "His Holiness, Pie Popus."

Lowell Thomas once referred to Secretary of Interior Harold Ickes as "Mr. Iksie."

Bill Leonard introduced Joe E. Brown as "currently starving on Broadway in *Harry*."

Cross My Heart

Milton J. Cross, known to generations of American radio audiences as the host of major musical events, once announced a concert in this manner:

"Ladies and gentlemen, may we call your attention to an outstanding musical event to occur next Saturday night—the first in a new series of concerts by the NBC Symphony Orchestra"—slight pause for effect—"under the direction of Ortosco Torganini." He tried again, and the second time it came out as "Artosco Turanini."

Cross once tried to introduce a well-known radio orchestra called the A & P Gypsies, but it came out as the A & P Pipsies. Another time he described "The Prince of Pilsen" as "The Pill of Princeton."

In perhaps the most famous of his bloopers, Milton Cross announced a newsbreak on one of his Texaco opera broadcasts as follows: "And now, stay stewed for the nudes."

Hello, Dr. Spooner . . .

Andre Baruch introduced *Your Army Service Program* by announcing: "And now, the orchestra, with Warrant Officer Edward Sadowsky seducting."

On another occasion Baruch welcomed his listeners with:

"Good evening ladies and gentlemen of the audio radiance."

Frank Knight, a weatherman who might have provided the model for Ted Knight on *The Mary Tyler Moore Show*, once predicted:

"The weather: tomorrow rowdy, followed by clain."

John Reed King referred to the Lindsay-Crouse hit on Broadway, *Life with Father*, as "the perennial Lice with Father."

Burl Ives was being presented with an award by Russel Crouse:

"I want to thank Brussel Rouse," he said.

Jerry Lawrence once explained: "When the King and Queen arrive you will hear a twenty-one-sun galoot."

During World War II Larry Lesueur reported that an admiral's comment about war material being sent to the Soviet Union under the lend-lease program created "a teapest tempot." Ed Murrow cabled from London to Paul White back at CBS:

"Please purchase suitably inscribed, old-fashioned, enameled, single-handed teapest tempot and present it to Lesueur on behalf of his admirers in Columbia's London office."

. . . and Mrs. Malaprop

R. H. Wood was the engineer in charge of BBC's Outside Broadcasts for three decades (1935–64), handling such events as royal weddings, coronations, and state funerals. He was a favorite with King George VI, who insisted on his presence at his annual Christmas broadcast. The king was usually nervous because of his stammer, but Wood's quiet, confidential manner put him at ease.

Wood was famous for his malapropisms, of which his colleague Brian Johnston preserved the following examples:

"He ran out of the room like a house on fire."

"Now he's buttered his bread he must lie on it."

"He's happy as a sandbag."

"He puts his finger into every tart."

People Who Ride Elephants Have Long Memory

One of the more difficult royal events Godfrey Talbot covered for the BBC was Queen Elizabeth's visit to India, when he tried to describe the journey down to the Ganges on elephant back. He kept slipping from his perch on the pachyderm, which made him punctuate his sentences with unconscious exclamations of "Whoops!" every minute or so. But Talbot's biggest slip came when he narrated:

"In a comfortable howdah high on the front elephant, Her Majesty, head of the Commonwealth, sits *entwined* with the Maharajah."

"Never since that day," Talbot wrote in his memoirs, "have I wittingly used the verb 'enshrine' in a broadcast, for fear of getting it wrong as I did then."

Royal Visit

Covering the progress of George VI and Queen Elizabeth across Canada was a major broadcasting event in 1939. Things went fairly smoothly until the royal couple arrived in Winnipeg, where they were greeted by MacKenzie King, the Canadian prime minister, and the city's mayor, whose name happened to be Queen.

"Here comes the Royal Family now," the announcer narrated breathlessly on CBC radio, "the automobile has now stopped, a member of the RCMP is opening the car door—oh, there's the King—he's stepping out, followed by Her Majesty Queen Elizabeth, nattily attired in a silver coat. Mr. King is now shaking hands with the King and introducing Mr. Queen to the King and Queen and then Mrs. Queen to the Queen and King. They are now proceeding up the steps to the well-decorated City Hall, the King and Mr. King together with the Queen being escorted by Mrs. Queen. The King has now stopped and said something to Mrs. Queen and goes to Mrs. Queen and the Queen and Mr. King and the Queen laughed jovially. The King leaves Mr. King and goes to Mrs. Queen and the Queen and Mr. King follow behind. The King . . ." Suddenly, the announcer fell silent, realizing that listeners might be getting a bit confused. When he came back on the air, he resumed with a new strategy: ". . . the Mayor and the King are now passing through the door of the City Hall followed by the Queen and the Mayor's wife . . ."

Fortunately MacKenzie King was a bachelor, and so there was no Mrs. King.

What's My Name?

G. W. "Johnny" Johnstone, reporting on the maiden voyage of the *Queen Mary*, began his broadcast: "Good afternoon, ladies and Johnstone."

Sterling North, the *New York Post*'s literary critic, who had a program on WHN, *Books on Trial*, signed off one evening with "This is Sterling Night signing off and saying good north and pleasant reading."

Whose Side Are We On?

During an episode of *Nick Carter* on Mutual Broadcasting, one German general was saying: "We are surrounded on all sides by the enemy—they come from the left, from the right—from the east, west, north, and south—and we are without food and water." To which his aide was supposed to respond: "Is it that bad?" Instead he muttered: "Is that bad?"

Clean and Unclear

One of the models for Revlon's "Clean and Clear" commercials studied long and hard to say her one line: "Soap on my face? Never!" When her moment came, on live television, she told 55 million viewers:
"Face on my soap? Never!"

The Blooper Man

One of the memorable snafus on early television had Betty Furness trying in vain to open a frigidaire; she keeps talking while tugging with small violent movements at the stubborn door, until a technician crawls to the fridge and springs it open. According to Kermit Schafer, this historic event took place not on *Candid Camera*, as it is widely believed, but on *The Rube Goldberg Show* that he produced in New York.

The incident gave Schafer an idea, and he went on to a self-created career of producing books, records, films, and shows about bloopers—all the little or big things that could and did go wrong with live television. With the advent of taped shows, it looked like the end of the bonanza, but then Schafer found a new gold mine in out-takes. "Performers feel they can take greater liberties," he said. "They feel that if they are no good or get tongue-tied, it can always be edited out."
When Burt Prelutsky once asked Schafer how he wished his epitaph to run, Schafer summed up his credo: "To forgive is human; to err, divine."

You Know What I Mean

The host of the long-running *Ed Sullivan Show* was famous for his fluffs, though it is now thought that his lapses of memory were due to arteriosclerosis. Following are three of his classic blunders.

Ed Sullivan once brought on stage a group of New Zealand natives and introduced them: "Please welcome the fierce Maori tribe from New England."

One Easter Sunday Sullivan had singer Sergio Franchi on his show to sing "Ave Maria." But the emcee could not pronounce the singer's name, and kept flubbing trying different versions ranging from the geographically inaccurate Frenchy to the Americanized Franchise. Finally he gave up and said:
   "How about it for 'Ave Maria'?"

Once night Ed Sullivan finished his program with a public service announcement that did not quite come out the right way: "And now a word about tuberculosis . . . Good night everybody, and please help stamp out TV!"

Wearing Freudian Slips

Ralph Edwards: "And here is one of radio's most charming and lovely young sinners."

Mary Margaret McBride: "A lot of things you are supposed to eat, you just don't like . . . especially children."

Jimmy Carter's inauguration happened on a chilly January day in 1977. Bob Schieffer, CBS correspondent, must have been thinking warm when he reported: "The Supreme Court justices have robes on but underneath they're wearing thermonuclear underwear."

The Myth of the Little Bastards

Many people will remember a few years back President Reagan's famous testing of the microphone before one of his regular radio broadcasts: "One, two, three . . . we begin bombing in five minutes." The mike was "live" and the message went out all over the country, sending a blast of chill into the world.
   Ronald Reagan, the former broadcaster, would certainly have known the most famous incident with an open mike in the early history of radio, involving "Uncle Don" Carney, who after one of his popular children's programs, was supposed to

have said something distinctly unavuncular over the air. Interestingly, this most famous of episodes from the early years of radio may be pure folklore. Bill Treadwell, a writer on the show and "Uncle Don's" biographer, quotes this earwitness account by the noted critic Oliver M. Sayler:

"It was back in the winter of 1928–29. The station was WOR. I was in the fifth year of my weekly book and play review, *Footlight and Lamplight*. One of Uncle Don's programs for children immediately preceded my time on the air. But I had no contact with it except when, occasionally, other studios were occupied and I was asked to broadcast in the studio he had used.

"On this particular occasion, I was to follow Uncle Don on the spot, and I was standing by in his studio, waiting for the late Floyd Neal to sign him off, give the station break and introduce me. Uncle Don twittered his usual cheery wind-up, and then, not realizing that I was to follow on the same microphone, and thinking he was off the air, blurted out: 'There, I hope that'll hold the little bastards!'

"Well, he wasn't off the air! The nation-wide reaction to his blunt statement raised a furor that impaired his celebrated program. Only after a 10-year atonement and the spending of a fortune on his part to conduct a children's entertainment concession at the N.Y. World's Fair in 1939, did he manage to return fully to radio's good graces. But he never again achieved his former vogue."

In fact, "Uncle Don" lasted uninterrupted for 23 years until 1949, when the old vaudevillian began to succumb to wine, women, and his own boredom. But Bill Treadwell's reflection on Sayler's account tells much about the power of myth over fact:

"If Don had really said this, he would surely have been removed from the air (which he definitely wasn't). . . . The story has been told so often even his close friends believe it, and why not? If you knew Don the way I did personally . . . you couldn't think of anyone more likely to make the remark than Uncle Don. A quip like that was like breathing to the guy. He should have said it. But the craziest quirk of his whole life was that he didn't!"

Happenings

Producers discovered early that if humans can make mistakes that are funny on television, animals can do it even better, and the audience forgives them for it. Herb Sanford, producer of The Garry Moore Show, *recalled in his book about those days some moments with Ivan Sanderson, who was the resident zoologist— and zookeeper—on the program.*

Ivan's appearances frequently developed into "happenings." One of these occurred with the African marabou stork, six feet in height with a seven-foot wingspread, brooding, inscrutable. Ivan and Garry stood at the footlights, the stork between them. Ivan remarked that the huge beak could bite through a plank of wood in a second. This explained why he and Garry were holding garbage-can lids as shields in case the stork decided to improvise. But, Ivan added, reassuring the audience, the stork could not take off from a standing position. Like an airplane, it needed a runway. Immediately, as if on cue, the stork took off, flew out over the audience, headed straight for the camera in front of the control room, and landed on the shoulders of a lady in the back of the house.

Pandemonium took over. Some of the audience began rushing for the exits. Those more sensible remained still. In the uproar, Ivan sprinted to the last row, which was unoccupied in order to accommodate camera platforms. With the help of his assistant. Eddie Schoenberger, Ivan managed to grab hold of the beak and haul the stork out of the audience and back on stage. The exodus that had begun was halted. All of this was caught by the cameras. It was manifestly unrehearsed.

Marcia Durant, who became our feature editor in 1953, recalls a moment of suspense backstage while a song number was on. Ivan's spot had just been completed; Marcia's interview subject was next. Marcia notice that the stage-hands were quietly retreating to offstage positions. Then she saw the anaconda peering out of its box and beginning a slow advance. Fortunately, Ivan also saw it and the anaconda was secured.

Ivan's visits did not always bring imminent danger. When Ivan heard someone use the expression "more fun than a barrel of monkeys," he thought it an excellent idea. The next week he arrived with a large barrel filled with monkeys. Just one monkey could be unpredictable. A barrel of monkeys was overwhelming. When the lid was removed, they charged in every direction. Some of them escaped into the flies, up to the highest level. And there they stayed. It took patience and cunning on the part of the stagehands—and several days—to persuade those monkeys to come down. . . .

Marcia's liking for animals led her to some outsize characters and a few rough moments. A newspaper clipping from a viewer steered Marcia to S. L. Ditto, retired newspaperman living in Florida. Mr. Ditto was owner of Lord Lee, a dancing goat. According to the newspaper story, Mr. Ditto played the harmonica and Lord Lee did a buck-and-wing. After several phone conversations, a date was set for Mr. Ditto and Lord Lee to appear on the show.

They were due to arrive at Idlewild (now Kennedy) Airport on a Sunday afternoon. Marcia enlisted the help of a friend, Lorrain D'Essen. Lorrain was owner of Animal Talent Scouts and did the animal casting for Broadway shows. Lorrain knew how to handle every kind of animal; having her along would be a help in dealing with Lord Lee. Marcia and Lorrain set out in Lorrain's station wagon. "I shall never forget," says Marcia, "that hot Sunday, driving out to the airport with Lorrain. We got stuck in traffic, but we got there. Coming in was more of the same. Lord Lee hadn't done a really terrific dance at the airport for the news photographers, but I figured both he and Mr. Ditto were tired and hot."

At the Henry Hudson Hotel, which had accepted Mr. Ditto and Lord Lee as guests, Lord Lee smashed a big mirror in his room. "He saw himself in it and thought there was another billy goat in the room. Lorrain and I were plenty scared when this happened. We thought the goat might have gotten mirror shards in his head. We got the mess cleaned up—and it was much more than mirror fragments! I worried about the ghastly smell of goat dung floating down the hall. During all this, Ditto was next door, air-conditioned, talking to his buddy from the newspaper.

"Then there was rehearsal. We had to order something special as 'footing' for Lord Lee—so he could do his buck-and-wing on a sure surface. We went through all those last-minute hassles to get this. I was on stage, taking notes. Ditto started playing his harmonica and the goat lifted his front legs and reared up, then went back on all fours. Then a voice from the control room said, 'Tell me when we should get ready for the goat to dance, Marcia.' It was Schim, I think I really blushed bright red. Then I said, 'He just danced, Schim.' And that was rehearsal. I can't remember if the goat danced a little or at all on the show."

The Buck Never Stops

Paul Heller, who went on to produce the film *David and Lisa*, was working as a set designer on a television commercial for refrigerators. The concept revolved around showing a lot of ordinary fridges until the one with the sponsor's brand name would be revealed in a flood of lights and with a baby elephant poised on top of it. But when the director called for lights and camera, the only action that ensued came from the poor animal's bowels, which discharged all over the shiny new refrigerator.

"Cut!" the director said, and turned to Heller: "Paul, we're going to have to clean that up, Paul."

Heller looked at the mess and said, "I make sets. That fridge is a prop. It's up to the prop man."

"That's not props either," a voice was heard from the back of the stage. "It's makeup."

"No way!" the makeup man yelled back at the prop man. "I only have to take *that* from established stars, not from elephants!"

Helpful Suggestion

To help newsmen with the pronounciation of proper names, suggestions are sometimes made on the TelePrompTer in the margin of the text. During a breaking story, these are occasionally incorporated into the text, providing landmines for announcers. Paul Moyer, anchorman of the local KABC television news in Los Angeles, was trying to report once on a statement from Admiral William Crowe, and said:

"The Chairman of the Joint Chiefs of Staff, rhymes-with-how, just announced . . ."

Tragedy

Christine Lund, anchorwoman on the KABC television station in Los Angeles, was once reporting a tragedy in Watts: children playing with matches in a closed garage were burned to death.

"We have a most tragic story," she intoned solemnly, "of some children who were playing with that most dangerous of toys—watches."

As the line sank in, the technical director's voice came over the P.A.:

"Christine, we've all taken off our watches here in the booth."

Grinlock

Bill Keene, who has been giving weather and traffic reports on KNX radio for thirty-five years, is a Los Angeles institution. He recently received a star on the Hollywood Walk of Fame. Part of Bill Keene's charm and fame lies in his endless wordplays and proprietary geography. In Keene's vocabulary, the vast freeway interchange just east of downtown L.A. is called "Malfunction Junction"; "Commerce Creep" occurs near the suburban town of Commerce where drivers slow down (forming "a gawkers' block") to look at the Assyrian-style landmark that used to be a tire factory; "Poop-out Hill" is the Sepulveda Pass on the 405 freeway, while "Thou Shalt Not Pass" refers to the creeping Hollywood freeway through the Cahuenga Pass.

Given the frustrations of driving in Los Angeles, Bill Keene makes one smile, and sometimes laugh outright, with his offhand puns. On one now famous occasion, when the California Highway Patrol arrived on the scene of a load of tuna spilled across several lanes of the Santa Ana freeway, Keene remarked:

"You've got fish and CHiPs there."

It's a Wonderful Town

Norman Brokenshire once ran short of material in his early days on WJZ. "Ladies and gentlemen," he announced, "the sounds of New York." Then he simply raised a window in the studio, and stuck the microphone out.

CHAPTER 4

News on the March

Politics As Usual

Norman Brokenshire assisted Major Andrew White in covering the tumultous Democratic convention in 1924, which began at the old Madison Square Garden on June 24 and lasted until July 10, when John W. Davis was finally elected on the 103rd ballot.

The proceedings leading up to each ballot were so fractious that warring parties engaged each other physically in floor fights. Major White happened to be away on a lunch break, and found young Brokenshire in an animated blow-by-blow description of the fighting. White quickly grabbed the microphone from his assistant, explaining afterward that WJZ got permission to broadcast the convention only if nothing would be reported that could cast the Democrats in an unfavorable light.

Why Adlai Lost

Co-anchoring the 1952 Democratic convention, Eric Sevareid and Walter Cronkite got their advance copies of Adlai Stevenson's acceptance speech just before his nomination.

"I was thumbing through my copy," Cronkite recalled more than once, "and I thought it was quite a speech, assuming that Eric was thinking the same. Then Eric threw the speech down and kind of grumbled to himself. 'What's the matter, Eric?' I said to him, 'Don't you like the speech?' And Sevareid looked up and said, 'I don't think I can care very much for a presidential candidate who writes better than I do!'"

Walter Cronkite enjoyed telling this story so much that Sevareid complained that he would never let him forget that episode.

"Oh, I don't know," Cronkite replied with glee, "I only mention it once in four years."

A Hard Day's Night

It was President John F. Kennedy's assassination on November 22, 1963, that established television as the most powerful and immediate medium for covering a breaking news story. Walter Cronkite, uncharacteristically in his shirt sleeves and his voice breaking with emotion, announced the president's death at 2:33 P.M. in the afternoon. The veteran newsman had tears in his eyes. Finally relieved from his post, Cronkite was trying to phone his wife, when the switchboard put through another call.

"May I have the news department of CBS?" a woman asked. Cronkite told her she had the right number.

"Well, I think it's absolutely criminal for CBS to have that man Cronkite on the air at a time like this, when everybody knows that he hates the Kennedys. But there he is, in shirt sleeves, crying his crocodile tears."

The emotion of three hours on the air amidst a great national tragedy had taken its toll, and the normally genial and controlled newscaster yelled into the receiver, before slamming it down:

"Madam, this is Walter Cronkite and you're a goddamn idiot!"

Silent Protest

For decades CBS was the preeminent network because of its news service. This was due in large part to William Paley's willingness, in the early days, to spend any amount to build up the news division. When he began in broadcasting, NBC was in first place by far, with the vast resources of RCA behind it. Paley had heard that Ramsay MacDonald, the British prime minister, would address the American people during his 1929 visit to the United States exclusively through NBC, and he protested loudly. When nobody listened, Paley ordered the whole CBS network off the air during the entire broadcast of MacDonald's speech. He lost quite a bit of money, but his silent gesture spoke louder than words.

The Chairman

William S. Paley had a genuine commitment to news, though this did not always benefit CBS. Despite innumerable changes of names, format, and personnel, the daily morning news program on CBS ran a poor third to NBC's *Today Show* and later to *Good Morning America* on ABC. Unlike the entertainment programming employed successfully by the competition, the *CBS Morning News* remained a magazine program built around the news. However, no major changes could be made, because Paley watched the program every morning and took an unusual personal interest in it. He often called Dick Salant, long-time president of the news division, telling him how a picture should hang on the set, or when commercials ought to start. Salant realized later that the chairman watched the program every morning religiously from bed, while his valet served him breakfast.

William Paley had made a modest investment from his family's cigar company in a string of struggling radio stations and built his media empire into one of the great fortunes. He had grown so rich that his mother was at a loss what to get him for his birthday. One year she gave him a valet with a note: "This is to remind you that you should never do for yourself what others can do for you."

Long String

For several decades CBS News was in a class of its own, and its reputation intimidated newsmen. In the mid-sixties, William Small, legendary head of the CBS Bureau in Washington, tried to recruit Douglas Kiker, who had been reporter for the *New York Herald Tribune* until its demise. But Kiker ignored the feeler and went to work for NBC instead, where he had a distinguished career until his death in 1991. Small still wanted to know why he did not give it a try at CBS, even if only to boost his negotiating stand with NBC.

"I looked up your staff in Washington," Kiker explained, "and I said: 'If I went to work for them, and the first string was wiped out in an airplane crash, and the second string died of heart failure, I'd still have trouble getting to be first string.'"

Cast No News Before Its Time

It was ITV, Britain's independent television, launched in 1955, that created the modern newscast, based on strong personalities and visual images instead of words. Before ITN, the BBC did not really know what to do about television news, even though it had done a remote broadcast in 1938 showing Neville Chamberlain wave the useless piece of paper he had brought back from Munich.

After television service was resumed in 1946, no newsreels were shown at all for a couple of years, on the express orders of Sir William Haley, the director-general, who believed that visual presentation "would prejudice all sorts of values on which the BBC's great reputation for news had been founded." Stranger still, newscasters were not shown on the air until 1955: instead, radio announcers read the news behind the forbidding image of the BBC clock. Even after competition from ITN forced the BBC to put newscasters on the air, for quite a while they remained nameless.

As newsreels were gradually introduced on the BBC, they were so stale that the entire newscast would be repeated twice weekly. But even that seemed fresh in contrast to the Fourteen-Day Rule, which outlawed discussion of any government policy or pending legislation for at least two weeks, so as not to prejudice discussion in the House of Commons. Hard to believe, but in 1950 the BBC failed to mention for several weeks the fact that a general election campaign had been under way; it was revealed to viewers only after the polls had closed.

In 1953, Churchill tried to prevent coverage of Queen Elizabeth's coronation from inside Westminster Abbey. (The BBC had broadcast the outdoor pomp of George VI's coronation as far back as 1937.) He lobbied the Archbishop of Canterbury and the Duke of Norfolk, in charge of the ceremonies, and finally told the sovereign that he and his cabinet were against televising the proceedings. "Rubbish," the young queen was reported to have told the elder statesman: "All this expense and everything done for this great occasion, it should be seen by people everywhere."

Truth in Broadcasting

Covering Queen Elizabeth II's 1977 Silver Jubilee for *The Today Show*, anchor Tom Brokaw speculated aloud that the secret of her enduring popularity "is simply that she has never once spoken to a journalist."

Cult

The cult of Edward R. Murrow, as the father of American broadcast news, was well established by the early fifties. It embarrassed Murrow, who was a shy and modest man. One of his radio newswriters, Jesse Zousmer, announced a "Murrow Ain't God Club" to puncture some of the myth. Some colleagues, whether out of fun or jealousy, began to join. The club suddenly died when Zousmer received a sincere letter from Murrow himself asking if he could become a member.

Broadcast News

"Broadcasters suffer from perfumed tonsils," H. L. Mencken once observed on the prevalence of pompous and self-conscious voices in early radio at the cost of what was being said. In recruiting William Shirer to be the Berlin correspondent for CBS News in the late thirties, Edward R. Murrow was looking for a good reporter, but his bosses, Paul White and Ed Klauber, wanted to hear a good voice. Shirer's was high-pitched and would crack when he became excited. Murrow won his battle to hire Shirer, and articulated a principle that seems to have been lost in the later development of American broadcasting:

"We can't risk using men with pear-shaped tones or pretty faces who know nothing about the world of reporting," Murrow told his superiors. "We must have highly competent professionals. I can teach a brain how to broadcast but I can't teach a pear-shaped voice how to think and write."

Never Mind

Like many radio newsmen, Edward R. Murrow was slow to warm to television, which he believed was good only for covering wars and sports. "TV is action," he once said, "and it is mindless." Not long afterward, he became one of the greatest names in early American television broadcasting.

The Man in the White Hat

Following the historic television broadcast of *See It Now* on March 9, 1954, in which Edward R. Murrow exposed Senator Joseph McCarthy and McCarthyism, the power elite in Washington held its collective breath to see what effect it would have on a government and nation which had been cowed into silence by the unfounded slanders from the junior senator from Wisconsin. Soon afterward, at a Gridiron Club dinner attended by the press corps, President Eisenhower came over to Murrow's table and playfully ran his hand up and down the newsman's back. As everybody watched, Eisenhower said loudly:

"Just wanted to see if there were any knives sticking in there."

There was relief, applause, and laughter. Murrow shook hands with the chief executive, whom he knew well from the war, and said equally loudly:

"From here on in, it's up to you, Mr. President."

Although Edward R. Murrow had the tacit backing of William S. Paley for his controversial television documentaries, the top CBS brass and the Aluminum Company of America, which sponsored *See It Now*, distanced themselves from the attack on McCarthy by refusing to look at any script or footage in advance. As the airing date approached, senior network executives went out of their way to avoid even bumping into Murrow.

After the overwhelmingly positive response to the program, one of them expressed the attitude of the top CBS brass with the following congratulation:

"Great show, Ed. Sorry you did it."

The Man and His Clothes

Charles Collingwood, one of the legendary CBS correspondents, hailed from North Dakota, but had gone to Oxford as a Rhodes scholar, and was a bit of a dandy. Morley Safer, who as a young reporter modeled himself partly on Collingwood, was much amused upon his first visit to Paris as a CBS newsman that the writer Janet Flanner inquired about his colleague with, "And how is the Duke these days?"

It was Edward R. Murrow who had recruited the Duke of Collingwood to cover many major battles of World War II, including the Normandy landing. Almost a year after Murrow had hired the young newsman, he confessed over drinks that at first meeting he had almost decided against him.

"Is it something I said?" Collingwood asked.

"No, Charlie," Murrow dragged on his cigarette, "but when you walked in wearing those godawful loud Argyle socks, I wondered if you were really right for us."

Harry Reasoner told Gary Paul Gates, author of *Air Time*, that Collingwood's mien and clothes made him feel "the discomfort of a man who's just noticed he has soup stains on his tie or that his fly is open." When Reasoner became a regular replacement for Walter Cronkite during his absences from the *CBS Evening News*, he wondered why the more senior Collingwood had been passed over. Somebody in the know told him:

"Because he's too fucking urbane."

How to Make Somebody Else's Career

Morley Safer, the urbane correspondent on *60 Minutes*, got his start at CBS by accident. He was a London-based reporter for the Canadian Broadcasting Corporation when the network's anchorman, Stanley Burke, decided—as so many other Canadians have done—to apply for a higher-paying job in American television. As a sample of his work, he sent to CBS News a tape of a year-end roundtable discussion by various CBC correspondents. The people at CBS News were more impressed by Morley Safer, and sent him an unsolicited job offer from New York.

Stanley Burke stayed on in Canada, though after a few years he left the CBC.

Safer made his reputation for realistic reporting from Vietnam, where CBS sent him in 1965. After he reported on the burning down of the Vietnamese village of Cam Ne by U.S. Marines, an infuriated Lyndon Johnson called CBS and asked if Morley Safer was a Communist.

"No," he was told, "Safer is a Canadian."

"Well," said the president, "I knew he wasn't an American."

Bad Night at Black Rock

When Charles Osgood, the elegant commentator at CBS News first came to work there, he was asked to substitute one Saturday evening for anchorman Roger Mudd. Unfortunately, almost everybody else was filling in that weekend: the executive producer, the director, the editor, the floor director, and the cameraman who worked regularly for the *CBS Evening News* were all away on holiday. "Somewhere along the line we had one too many substitutes that night," Osgood wrote in one of his recent collections of columns, *The Osgood Files*. After he would introduce a filmed item, invariably the wrong report was shown, or there were other pieces that he knew nothing about.

"But the worst part came," the witty and unflappable Osgood reminisced, "when I introduced the 'end piece,' a feature story that Hughes Rudd had done about raft racing on the Chatahoochie River. Again, when I finished the introduction, I turned to the monitor and, again, nothing happened. Then, through the glass window of the 'fishbowl,' I heard a loud and plaintive wail. 'What the **** is going on?' screamed the fill-in executive producer. I could hear him perfectly clearly, and so could half of America. The microphone on my tie clip was open. Standing in the control room watching this, with what I'm sure must have been great interest, was a delegation of visiting journalists from the People's Republic of China. They must have had a really great impression of American electronic journalism. The next Monday morning, sitting back at the radio desk where I belonged, I became aware of a presence standing quietly next to my desk. It was Richard Salant, the wise and gentle man who was then president of CBS News. He'd been waiting until I finished typing a sentence before bending over and inquiring softly: 'What the **** *was* going on?'"

Someone to Fear Over You

Don Hewitt, the producer of CBS's 60 *Minutes*, commands fear and loyalty in his employees. In 1959 Hewitt, then producer of the *CBS Evening News*, assigned Harry Reasoner to cover Nikita Khrushchev's visit to the United States. With great ingenuity Reasoner managed to slip through the tight security ring around the Soviet leader to get an exclusive interview. Afterward, a State Department contact told the reporter that he might have landed in serious trouble for breaching security.

"You don't understand," the mild-mannered Reasoner replied. "I'm more afraid of Don Hewitt than I am of the Secret Service."

Rather Agonistes

Dan Rather first attracted national attention for dramatic coverage of Hurricane Carla that devastated Galveston in 1961. As news director for KHOU, the CBS affiliate in Houston, Rather practiced what he called the "fuzz and wuz" approach to local news: lots of film depicting police (fuzz) and dead bodies (wuz). As a reporter he became so well known for his pushiness that his colleagues circulated a photograph that showed President Kennedy surrounded by newsmen, with Rather suspended horizontally above them.

Dan Rather broke the news that JFK had died 17 minutes ahead of the other networks, even though for several of those minutes there was no official confirmation. His scoop ultimately led to the most visible beat in journalism—the White House. There Rather became famous for his combative style. Once when his tough questioning during Nixon's reelection campaign drew cheers from the crowd, the President fixed Rather with his gaze and asked:

"Are you running for something?"

"No, sir," Rather hit back, "are you?"

President Nixon was campaigning on behalf of Republican candidates for Congress during the 1970 campaign. Once when he arrived with the press corps in tow to Tennessee he was chagrined to see signs that read: "WELCOME DAN RATHER."

Show Business

When Dan Rather became anchorman for CBS News, he made a great deal of the fact that he was also managing editor of the program. Fred Rothenberg, television columnist for the Associated Press, came to interview him at the offices of the news division, and Rather put on a great show: he discussed story ideas with producers, gave orders to researchers, checked scripts with writers. Watching this unusual burst of activity from the managing editor, producer Lane Venardos said to executive producer Howard Stringer:

"Just think: people are paying fifty dollars a seat for theatre tickets just ten blocks away, and we're getting it for free."

A Little Learning

At a time when the ratings for the *CBS Evening News* were falling behind, Dan Rather called together the whole staff to give them a pep talk.

"I only have one thing to say to all of you people," he finished his rousing speech, and paused for effect. "Syracuse 413. Read it."

Then the managing editor left his troops, somewhat puzzled, until one of the researchers, Toby Wertheim, traced the reference to Sir Edward Creasy's 19th-century volume, *Fifteen Decisive Battles of the World*, which was Rather's pet reading at the time. In 413 B.C. the city of Syracuse almost surrendered to the Athenians, until the badly divided inhabitants rallied and pulled a victory from almost certain defeat.

Dan Rather reached the pinnacle from humble roots in local news, and always felt outclassed by his worldly and intellectual colleagues at CBS. Anxious to improve himself, he sought and listened to advice, bought and read through the Great Books, and once told the *New York Times* that on the advice of his "guru," Eric Sevareid, he always used to carry on his person Strunk and White's *The Elements of Style*, the classic primer on good writing. But when Sevareid was questioned about this story, he merely said:

"I've never read the book in my life."

Once Dan Rather was waiting to catch a plane from London to New York with Howard Stringer, the Oxford-educated executive producer of CBS News. Rather was drawn to an edition of *The Decline and Fall of the Roman Empire*, the 18th-century classic by the English historian Edward Gibbon.

"You read that in the original Latin, I suppose," Rather queried Stringer.

Anchors Aweigh

Ted Turner got into the news business without knowing too much about it; he readily confesses that before launching his Cable News Network in 1980 he had not watched more than a hundred hours of television news in his whole life. But Turner knew about stars, and asked Reece Schonfeld, the first president of CNN, to hire as chief anchor someone who would put the fledgling network on the map at once:

"Who's the biggest guy we could get?"

"Well, probably Dan Rather," Schonfeld suggested.

"Who's Dan Rather?" asked Turner.

Patiently, Schonfeld explained to the man who wanted to launch the first all-news television network about Rather, the best-known newscaster in America at the time.

"How much could we get him for?" Turner persisted.

"Oh, a million dollars a year," Schonfeld guessed.

"A million bucks just to read the news?" Turner marveled. "I just offered Pete Rose a million dollars to play for the Braves, and he only works half the year."

CNN never competed with the networks or even major local stations in paying extravagant salaries to anchormen and women, believing that the news itself should be the star. According to legend, one anchorman went to see Ted Turner about a raise.

"If I shake that tree," the boss pointed to a tree outside his window, "I bet I'd see anchors falling all over."

Showdown

Peter Jennings did not want to be network anchorman; he was quite happy as ABC's London correspondent. But once he accepted the job, he took seriously both the responsibilities and the power that goes with being managing editor. He had a showdown with Bill Lord, executive producer of the evening newscast, over reporting a story about Prince Charles and Princess Di's marital troubles. Jennings, the son of a famous Canadian newsman, Charles Jennings, thought this was merely gossip. When Lord pulled rank and insisted on the piece, Jennings introduced it in his own words: "And finally this evening, gossip at the highest level."

Bill Lord was heard raging in the control booth but felt vindicated afterwards because the other networks also ran the item. Jennings had the last word.

"Any anchorman who thinks he's always right is a fool," he declared. Then he added: "Any executive producer who thinks he is always right is a fool—for a shorter period of time."

A few weeks later Bill Lord was no longer executive producer of ABC News.

The Boss

The news business can be all consuming and takes its toll on the reporters' health and family life. Gordon Manning, one of the legendary executives in American television news, once complained to his boss at CBS, Fred Friendly, that constant travel was ruining his home life.

"If I don't start spending more time with the kids and Edna, I'm afraid she's going to divorce me."

Friendly looked at his haggard employee and said:

"I can live with that."

Like many in the business, Manning was a workaholic. He was in Las Vegas once for a conference and staying at Caesar's Palace. He called up Osborn Elliot at seven o'clock in the morning, just before they were to meet for breakfast.

"I just thought you might be wondering why they have mirrors on the ceilings of all the bedrooms here."

"Why is it?" Elliot asked, lost for an answer.

"That's so you can shave in bed."

When Lee Townsend, one of the editors of the *CBS Evening News*, returned to work after only a few weeks following a heart attack, his boss was suprised to see him.

"Well, Lee, you don't look a day older."

"I almost wasn't," Townsend replied.

Good-night, Chet!

Even before radio reached his home in rural Montana, Chet Huntley was gaining his first experience at announcing. The popular television newscaster's father worked a ranch but also possessed a license as a railroad telegrapher. The World Series and other exceptional sporting events were relayed play by play via the telegraph; Pat Huntley would decode the Morse and then young Chet "announced" the result to the small knot of people who gathered around the depot. He would always remember his debut, "broadcasting" the Dempsey–Firpo fight.

Chet Huntley possessed the rugged good looks of a Western star, but he did not become nationally known until he was in his mid-forties, when NBC news chief Frank Reuven teamed him with the acerbic David Brinkley. They anchored the political conventions in the summer of 1956 and began their regular *Huntley–Brinkley Report* on October 29, 1956. Despite the fact that it was the busiest week for news since World War II—with the Suez crisis aggravated by the Hungarian revolution—the ratings were dismal at first. But by the end of the decade, the *Huntley–Brinkley Report* had climbed to the top among the evening newscasts. Huntley and Brinkley became household names, their sign-off a national catchword. At the Kennedy inaugural party, Frank Sinatra and Milton Berle sang new words to a popular hit, "Love and Marriage":

Huntley, Brinkley,
Huntley, Brinkley:
One is glum.
The other twinkly.

No News Is Bad News

Finding it curious that the network newscasts were always the same length, David Brinkley startled NBC producers and executives one slow news day when he announced on the *Huntley-Brinkley Report*: "There's no news tonight in Washington." Later that week he received a letter from an old lady, chiding him:

"What d'you mean there's no news tonight in Washington? Don't you know the local Safeway supermarket is on strike?"

Very Special People

Barbara Walters became the first woman to co-anchor a major network news program in America, and many women reporters look up to her as a pioneer. Apart from her lack of journalistic background, Walters faced the biggest challenge from her co-anchor on the *ABC Evening News*. Harry Reasoner was a proud veteran from the great tradition of CBS News, and he called her hiring a stunt. Reasoner seemed, in the words of columnist Roger Rosenblatt, "as comfortable with Walters as a governor under indictment." In media circles the pair quickly became known as "Price and Pride."

Soon after her debut on the *Evening News*, Walters began her celebrity interviews, which helped to turn her also into a celebrity. The first *Barbara Walters Special* in December 1976 featured President-elect Jimmy Carter and his wife Rosalyn at home. Testing her signature technique of prying ("I don't know how to ask you this, so I'll just ask it"), Walters questioned the Carters whether they slept in double or twin beds. Innocent enough by today's standards, the question shocked many Americans at the time; perhaps unwittingly, Barbara Walters paved the way to the tabloid television of today. (For a spell, schlockmeister Geraldo Rivera worked with Walters on the program 20/20). But what upset other journalists most about the Carter interview was the ending when Walters begged him: "Be wise with us, Governor, be good to us." Marlene Sanders, who would soon quit ABC, said she felt like throwing up, and Morley Safer asked on radio: "What right does any reporter have to issue such a benediction? . . . It is as if Mr. Carter had just become Louis XIV and, without Pope Barbara's admonition, he might be dumb with us and mean to us."

Despite derision and constant attacks, Barbara Walters worked hard to get her stories. She could be very persistent, cajoling, pleading, and pushing for an important interview. During the Iran–Contra affair, she happened to be on the same flight with Leonard Garment, who was the attorney representing former national security adviser Robert McFarlane. Walters hovered over Garment and fired questions at him, which soon drew the attention of the whole cabin. "You think this is Barbara Walters?" Garment appealed to them. "It's really just a woman wearing a mask." As the plane was preparing to land, a flight attendant ordered Walters back to her seat. When she still wouldn't move, the concerned attorney quipped:

"Tell the pilot he'll have to circle three times. Barbara Walters is working."

Scoop and Poop

Among her journalistic feats, perhaps the highpoint came when Barbara Walters managed to get the first joint interview with Menachem Begin and Anwar Sadat in November 1977. For this scoop, she received the Hubert H. Humphrey Freedom Prize from the Anti-Defamation League of the B'nai B'rith. At the banquet where she was honored along with Walter Cronkite and John Chancellor for their diplomatic efforts toward peace, Cronkite demurred:

"The intent of none of us was to further the cause of peace. It was to get the story."

Post Haste

With today's instant satellite communications, it is instructive to read about the way a presidential speech was reported on the eve of the Second World War. On August 25, 1939, President Roosevelt made an appeal for peace to Hitler and President Benes of Czechoslovakia. A. A. Schechter, in charge of News and Special Events at the National Broadcasting Network, narrates how he got the news out that day to the American people.

We had been notified—as had all other news agencies—that a statement giving the text of the President's plea would be issued by the State Department.

At once men from our Washington office were dispatched to places where they could do the most good. It is against regulations to run lines for broadcasting into the State Department Building, so we had to protect ourselves in other ways.

I instructed one man to fill his pockets with nickels and take over one of the building's telephone pay-stations. His instructions were to stay in the booth and keep feeding nickels into the coin-box. He was to keep the line open so that if necessary we could jam the President's message through with no delays over that wire through the Washington office.

A second man was instructed to take up a position in a taxi as close as possible to a door leading from the State Department building. The door of the cab was to be open and the motor running.

Two other men were instructed to attend the press conference at which the President's message was to be released. Each was to grab a handful of releases and run like hell. One release was to be rushed to the man in the telephone booth, another to the man in the taxi.

As soon as Press Officer McDermott of the State Department appeared with an armful of releases, our men in the pressroom did their stuff. One sprinted with a copy to the man in the booth, another rushed the release to the man in the taxi. While the car headed for our Washington studios, the man in the booth said to the person at the other end of the phone in our capital newsroom, "I have the President's message." Whereupon Washington asked New York, over another telephone line that had been kept open so that a call could be put through immediately, "Give us the air."

I immediately ordered the air "thrown" to Washington.

Only five or six seconds elapsed between Washington's request for the air and the announcer's, "We now take you to Washington," followed by, "Ladies and gentlemen of the radio audience, the President of the United States has just released a message he has sent to Chancellor Hitler and President Benes. We will now read the text of the message."

The man in the booth was told to "come in." He started reading the message, picking his pauses with great care so that the announcer in the studio, who was repeating his words, could do so with a minimum of raggedness. For the first few minutes the broadcast was slow. Then our man in the taxi arrived with a copy of the text and the announcer was able to take the occasional suggestion of jerkiness out of his delivery and speed up his reading job.

As a result of the arrangements we had made we were fourteen minutes ahead of the rest of the field in giving the full text of the President's message to the listening public.

Eyewitless News

The hours of news broadcast on local television stations throughout America typically proceed from gruesome tales of shootings, rape, and arson to a magazine format of soft features, interspersed with sports and weather. The whole is held together by anchormen and women engaging in Happy Talk, a phrase originally coined by *Variety*'s Chicago correspondent Morry Roth to describe the nightly antics of the news team on WLS. The format, introduced there in the late sixties, spread to the *Eyewitness News* programs of the ABC-owned stations, and eventually found universal acceptance on most local stations, blurring the distinction between news and entertainment.

The basic principle behind Happy Talk is to fill the ether with constant commentary, no matter how inappropriate, and to make the viewer less uncomfortable with the bearers of a constant stream of ghastly news. In one famous instance, weatherman Tex Antoine was fired from WABC in New York, when following a news item about the raping of an eight-year-old girl, he commented:

"With rape so predominant in the news lately, it is well to remember the words of Confucius: 'If rape is inevitable, lie back and enjoy it.'"

Media critic Ron Powers wrote that those words, "presumably spoken in the lighthearted spirit that flavors all Eyewitness newscasts, were of such blatantly bad taste that they even exceeded ABC's permissive notions of propriety—a spectacular achievement in itself."

Not Necessarily the Weather

Ron Powers also gives a brief example of Happy Talk from its originators in Chicago, an exchange between anchormen Joel Daly and Mike Nolan introducing a weather report by John Coleman:

DALY: Well, what kind of cat-and-mouse games do you have for us in the weather, John?

COLEMAN: I'd be willing to discuss the weather, Joel, if I knew that nursery rhyme, "Ding, dong, bell . . ."

DALY: "Pussy's in the well."

COLEMAN: Go on.

DALY: I don't remember the other . . .

COLEMAN: I never heard that nursery rhyme, did you, Mike?

NOLAN: Oh yeah, I heard it.

DALY: That's right. "Who put him in?"

COLEMAN: Who?

DALY: Little Johnny . . . Coleman . . . ! [General merriment.]

COLEMAN: Aw now, cut that out. Well, I'm sure we're not experts on nursery rhymes, but I am reasonably well informed meteorologically at this moment, and a one-word comment would be: YAH-HOOOOO!

Revolting Newsmen

Many a time viewers may wonder how television could serve up the pap that it does. Even more phenomenal, though, is the way broadcasters and performers can dispense endless amounts of stuff that they themselves do not believe. John

Bohannon, a popular American radio broadcaster, collected examples of rebellion in his little book called *Kickers*. A German newsman in Hamburg, unaware that his mike was still open, once concluded his reading of the evening news with, "All lies." Later the station tried to limit the damage by saying that the anchorman was only referring to the final news clip.

Another anchorman working for Nigerian television, in Lagos, controlled by the government, was reported to have thrown down his script in the middle of the newscast.

"I'm fed up with this false information. I can't continue with this stuff." And he walked out—to an unrecorded fate.

Network

Movies like Network *and* Broadcast News *have dealt with the pressures on-camera reporters face on national television. Linda Ellerbee, in her book* And So It Goes, *tells of when she made her jump from local news into the big time.*

. . . So many people who work for local television stations think they would rather work for a network. I was one of them. A network seemed saner. I didn't know any better then. Besides, going to the network was considered, as a rule, a step up. Network news was more serious than local news, that was understood.

Ellen Fleysher and I talked about that very thing one night about three months after I had left WCBS to go to work for NBC News in Washington, D.C. At the time, we were sitting at the bar of the Park Lane Hotel on Central Park South in New York City, a well-mannered place to have a drink. Ellen had come to work at WCBS sometime during the two years that I had worked there; Ellen still worked there. She still worked as a local reporter. I, on the other hand, had "gone network." We talked about my new job, my future, the important stories I would cover, the important people I would meet, the places I would see. No more five-alarm fires in the South Bronx for me. This was the real thing. The big enchilada. Network. Network television news. I may have smirked a bit.

The bartender came over to where we were. We ordered. He paused.

"Excuse me, but aren't you Ellen Fleysher from Channel Two?" She said she was. He looked at me, the network journalist.

"Hey—didn't you use to be on television, too?"

Women in Television

Just before the mid-term elections of 1974, Richard Salant, president of CBS News, escorted Leslie Stahl, then a novice correspondent, to the set that was being constructed for the coverage of election-night returns. Stahl had been nervous before her first major national exposure and Salant hoped that she would feel better after familiarizing herself with the lay of the land. They inspected the rough set where workers were busy putting in the various anchors' desks. Stahl noted the legendary names that were sketched on each post: Cronkite, Wallace, Rather, Mudd. And then her eyes fell on the desk where she would be sitting within a short while, which read, much to Salant's horror: "FEMALE."

Barbara Walters was fired from her television job at CBS in 1957. At one point she asked Don Hewitt about her chance for a career.

"Barbara, you're a marvelous girl," the powerful producer told her, "but stay out of TV."

Jane Pauley, who became the popular co-host of *The Today Show* on NBC, began her career in 1972 by applying to a local station in her hometown of Indianapolis which was advertising "for a female-type person" as a reporter.

We Are Still in Kansas, Dorothy

Christine Craft became a national celebrity when in 1983 she sued KMBC, a Kansas City television station owned by Metromedia, for sex discrimination. Management fired the thirty-seven-year-old anchorwoman, and explained their action with notable tact:

"The people of Kansas City don't like watching you anchor the news because you are too old, too unattractive, and you are not sufficiently deferential to men. We know it's silly, but you just don't hide your intelligence to make the guys look smarter."

When Craft's lawsuit required the station to back up these assertions, it was revealed that KMBC had hired a market research firm to conduct focus group discussions among the viewing public. One question addressed to an all-male group began:

"Let's be honest about this, she's a mutt, isn't she?"

One Who Overcame

In the middle of her trials, when she was making many appearances all over the country, Christine Craft went to Cleveland to be interviewed by Dorothy Fuld-heim, America's oldest anchorwoman, still going strong then at ninety. Craft was hoping, of course, to enlist support and draw inspiration from the legendary role model, but found Fuldheim the toughest interviewer she had experienced.

"Her attitude was one of 'Show me you're not just a crackpot, give me one good reason why I should be interested in your plight,'" Christine Craft wrote in a book about her *cause célèbre*. "The person who many would have presumed to be my staunchest ally was in fact the hardest to convince that I was telling the truth. She never gushed approval or support, but quietly in the tone of the questions she asked me, I sensed she knew I was fighting a battle she had won many times over. Her parting shot was lighthearted and encouraging: 'I've just signed a new three-year contract. I've just turned ninety. I call that optimism!' With a quick wink, she returned to her life's work of being simply smarter than anyone who would challenge her right to stay on the air."

Risqué News

Mary Alice Williams, one of the top reporters and anchors in the early days at CNN, recalled some of the hazards of doing live, non-stop broadcasting. Once Stuart Varney was stuck in traffic, and Williams was dragged from her dressing room to sit in for him. She managed to get curlers out of her hair, but when she tried to put on the mike, her dress fell apart in the front, and she was facing the camera in a black bra.

"They zoomed in extremely close while I was holding the dress up along with the mike," she told writer Hank Whittemore, "and I just couldn't stop laughing."

A few years later, Mary Alice Williams was reporting from the 1988 Republican convention in New Orleans when a wire ripped her skirt wide open. As she bent over in the anchor booth, Williams accidentally "mooned" thousands of delegates on the floor below. The Grand Old Party took it in stride.

Eschatology

In announcing the launching date of CNN on June 1, 1980, the world's 24-hour cable news network that would never shut down, Ted Turner told reporters:

"Barring satellite problems, we won't be signing off until the world ends. We'll be on, and we will cover the end of the world, live, and that will be our last event. We'll play the National Anthem only one time, on the first of June, and when the end of the world comes we'll play 'Nearer My God to Thee' before we sign off."

It's a Bomb

In June 1946, the four radio networks pooled their resources to cover the U.S. atomic bomb test on the Bikini atoll. The show business newspaper *Variety* ran the following review of the 50-minute program:

ZZZ ZZZ FFFTZ ZZZZM
ZZZ PFFFT ZZZZ ZZZZ
ZY PFTTT PFFFF ZZZZ.

CHAPTER 5

Under Duress

The Founder

Although the BBC has assiduously avoided the cult of any personality, during her formative years "Auntie" bore the unmistakable stamp of her "onlie begetter," John Reith. "A curious and stormy blend of Scottish Calvinism, snobbishness and self-assertion rumbled furiously in his powerful but not very coherent mind," wrote Malcolm Muggeridge about the founder, asserting that Lord Reith was perhaps "the only truly remarkable man ever to have been eminently concerned with it [the BBC]."

By his own confession, "born intolerant, reserved and aloof," Reith remained a lifelong teetotaler, who insisted that all announcers come to work in evening jackets, even though they could not be seen on radio. His favorite religious program, *The Epilogue*, closed down the BBC every evening. When the announcer was found necking with another employee on the premises, he was permanently banned from the program.

John Reith rarely interfered directly in specific programming decisions, but equally rarely did he miss an opportunity to inculcate his moral principles at the BBC. One day he visited the office of Captain C. A. Lewis, the deputy director of programs, with a copy of *The Times* in his hand.

"I see Daisy Kennedy's performing tonight," he observed.

"Oh, is she?" said Lewis, who was several weeks ahead in his schedule. "Anything wrong with that?"

"Haven't you seen the papers?" Reith went on. "She was divorced yesterday. We can't have a divorced woman performing."

On another occasion, Reith forced the resignation of his chief engineer, Captain Peter Eckersley, when he heard that his name would be mentioned in a divorce case.

Much later in life, Reith's extreme Puritanism mellowed somewhat. Martin Esslin, head of Radio Drama, remembers a telephone call from the founder, asking his help to obtain tickets for the Arts Club production of a risqué play that had been banned by the Lord Chamberlain in public theatres. Lord Reith was anxious to get the best seats up front.

In June 1938, John Reith resigned from the BBC at age forty-eight after sixteen years of service; he later called this the greatest mistake in his life. His complex and eccentric character is recalled in a story by Sir Robert Lusty, who had served on the Board of Governors of the BBC. In 1962 Lord Reith accepted an invitation to attend a Guildhall banquet to commemorate the fortieth anniversary of the BBC, on one condition: he would not have to make a speech. Sir Arthur fforde, chairman of the governing body, gave his assurances. On the day of the banquet, Reith telephoned the man in charge of the ceremonies.

"It is clearly understood," the founder asked gravely, "that I would not have to speak tonight?" The man confirmed it. There was a pregnant pause.

"But why," his lordship rumbled, "am I not allowed to speak tonight?"

Wuthering Height

The British Broadcasting Corporation, although accused frequently of being a tool of official policy, has a long history of conflict with various governments. During the General Strike of 1926, John Reith, the founding director-general, clashed with Churchill, then Chancellor of the Exchequer, who thought "it would be monstrous not to use such an instrument to the best possible advantage." Reith insisted on reporting the facts, which earned him the undying enmity also of the striking miners. "He behaved quite impartially between the strikers and the nation," Churchill railed against Reith much later in life to his doctor, Lord Moran: "I said he had no right to be impartial between the fire and the fire brigade."

Reith had left the BBC in 1938, and he was appointed the new minister of information by Neville Chamberlain. Churchill suspected both Reith and the

BBC to be working against the war effort, and as soon as he became prime minister, in May 1940, he removed him to a lesser portfolio. "There he stalks," Churchill once pointed at Reith, who was enormously tall, "that Wuthering Height"—an epithet that stuck for the rest of his life.

If Music Be the Food of War

Sir Winston Churchill was generally suspicious of the media, partly because he recognized its power for propaganda. During the war he advocated firm government control over the BBC, even in its musical programming. When challenged in the House of Commons, the prime minister argued: "I think we should have to retain a certain amount of power in the selection of the music. Very spirited renderings of 'Deutschland Über Alles' would hardly be permissible." (The German national anthem, of course, shares a tune with the great Anglican hymn which begins with the line: "Glorious things of thee are spoken, Zion, city of our God.")

But when Churchill was asked whether a pacifist musician should be banned from playing on the air, he replied with his usual common sense:

"I see no reason to suppose that the holding of pacifist views would make him play flat."

Every Man Has His Price

Sir Winston Churchill's broadcasts during World War II are often credited for inspiring the British with the necessary fortitude to resist Hitler's total conquest of Europe. Not long after he became prime minister, Churchill was preparing to address the nation. His driver had broken down somewhere and could not pick him up, so Churchill took a taxi to the BBC. On arrival, he asked the cabby if he could wait twenty minutes. Not recognizing the prime minister, the man said:

"Sorry, guv'nor, but I must hurry home to the missus to hear Mr. Churchill's broadcast. We wouldn't miss his speech for the world."

The prime minister was so pleased that he gave fifteen shillings for the three-shilling ride.

"To 'ell with Mr. Churchill," the cockney said with newfound enthusiasm, "I'll wait."

Arrant Pedantry

The British Foreign Office routinely checked the text of Churchill's wartime speeches before his broadcasts. In one of them, a zealous undersecretary questioned the prime minister's use of a preposition at the end of a sentence. Churchill, who won a few years later the Nobel Prize for Literature, was outraged at the query. He scribbled back a memorable memo to the civil servant:

"This is the kind of arrant pedantry up with which I won't put."

Fairy Tale

During the war Radio Eireann was touring various cities in the Emerald Isle with a Sunday-night quiz show called *Question Time.* One night, when the broadcast emanated from Belfast, the capital of Northern Ireland, quizmaster Joe Linnane asked one of the contestants: "Who's the world's best-known teller of fairy tales?" Instead of Hans Christian Andersen or the Grimm Brothers, the answer came back unexpectedly: "Winston Churchill."

The response brought down the house with cheers and laughter, but in Westminster angry questions were asked about Ireland's neutrality in the war. It would be some time before Radio Eireann was invited or allowed to visit Northern Ireland again.

Man in the Street

During the Second World War, Gilbert Harding was in charge of a program called *Meet John Londoner,* which the BBC made for American audiences, with the purpose of bringing the two allies closer together. Harding roamed around London with his colleague Stewart MacPherson in search of Londoners who might have views that would interest Americans. During one of Eleanor Roosevelt's visits to the GIs stationed in England, the broadcasters ambushed a man in the street (actually it was on Trafalgar Square) for his opinion about the First Lady's stay in London.

"We are proud and honored to have her here," the man replied in a fine cockney accent, "and to welcome her, and through her, her wonderful husband."

Encouraged by this unprompted outburst of perfect copy, Gilbert Harding asked:

"Have you seen Mrs. Roosevelt since she has been in London?"

"No," said the man in the street, "but I understand that her chief purpose in coming here is to have intercourse with the American troops."

Get Smart

The BBC was active on many fronts during World War II. There was unrelenting propaganda to keep the British spirit up in the face of the Nazi onslaught. There was also a secret war, which involved BBC personnel in various dirty tricks, such as monitoring radio and wireless messages between the German High Command and their field commanders, and using specific information, including names of individual officers, to broadcast disinformation or sow doubt about their spouses' fidelity back at home. Although British Intelligence cracked ciphers and inter- cepted messages with superb efficiency, the system occasionally broke down, as the Viennese-born Martin Esslin found.

Early in the war the Monitoring Service was evacuated to Evesham, a small town near Stratford-upon-Avon. I was with one unit, which monitored regular broad- casts from occupied Europe and Germany, and many of us were, of course, refugees and native speakers from those countries. There was also a highly secret unit at Evesham, working in separate huts, which none among us was permitted to enter, unless one was born and bred British, in the mistaken belief that real English gentlemen would never stoop to become spies. As we know, exactly the opposite turned out to be the case: only English gentlemen were found to have betrayed their country, and in fact, one of the most notorious, Guy Burgess, was working for the BBC Overseas Service at that time. Anyway, I later found out that this secret Y unit was monitoring flight communications and telephone conversa- tions from Europe.

Not long after the fall of Paris, in the summer of 1940, my boss came to me with a document and asked me to sign it, swearing under the Official Secrets Act that I would never divulge what I was about to hear. Then he asked me to listen very carefully to an intercepted telephone conversation which had been recorded on a wax cylinder. My task was to double-check the assertion by Y unit that the secret message contained important information about a German invasion of Portugal the following Thursday. I thought this was highly unlikely, with the Germans still on the northern side of the Pyrenees—they would first have to pass through the territory of their neutral ally, Franco's Spain—but I put on ear- phones and heard the following German conversation:

"Hallo, hallo, German News Service, Berlin, this is your correspondent in Lisbon. I will give you now my daily summary of the English press." The man's job consisted of reading the London newspapers that arrived daily from England by air. After his report, he asked:

"Please put me through to the administration." When they connected him, he asked for the pay clerk, and again they obliged him. "Listen," said the Lisbon correspondent, "I haven't received any salary for the past three months, and I'm down to my last penny. If I don't get paid by next Thursday, Germany will have to invade Portugal to get me out of debtors' prison."

I had great difficulties explaining to my superior that the gentlemen of Y unit predicted the invasion of Portugal from a joke that they did not understand.

Rude Interruption

Newsman Bert Silen was in Manila when the Japanese began their attack, on the same night as Pearl Harbor.

"Hello, NBC," he announced, "this is Bert Silen speaking from Manila. And this time, I've got a real scoop for you. Manila has just been bombed! In fact, right now it is being bombed! . . . Right now the moon is shining—absolutely full. It stands out like a mirror and it is no wonder that every bomber could pick out any spot around this part of Manila tonight. It wasn't the fault of the blackout, there isn't a light shining any place, but old man moon just wouldn't stay blacked out."

Silen kept talking on an open circuit to NBC until Japanese soldiers literally wrested the microphone from his hands, and dispatched the reporter to a concentration camp.

Later, General Douglas MacArthur was in the process of retaking the Philippines. His man in charge of communications, Abraham Schechter, knew about ongoing night raids to liberate the POW camps on the islands, and let it be known that Bert Silen, if and when found, should be brought immediately to his office. It was done, and within hours Silen began broadcasting on NBC with an appropriate opener that was somewhat less shopworn then:

"As I was saying the last time when I was so rudely interrupted . . ."

The Price of Freedom

Radio, which was largely developed during the First World War, became a victim of World War II, when legitimate fears arose that the enemy might exploit the medium for his own ends. In Weapons of Silence, *Theodore F. Koop, who worked in the Office of Censorship in Washington, describes the problems he faced.*

Security might be endangered by various types of programs—request numbers, quizzes, interviews—in fact, any broadcast in which persons other than trusted employees had a part. The possibility must always be considered that an enemy agent, by one means or another, could use a radio program to communicate a secret signal or code to his superiors at home.

If anyone doubts the feasibility of such an approach, he should be reminded of the test conducted by a Military Intelligence officer before the United States entered the war. He skillfully wove into a broadcast interview with Max Baer a coded reference to ship movements by a passing mention of Queen Elizabeth in the regal sense rather than as a liner. To the uninformed, the script was natural and innocent. When the officer disclosed the ease with which he had arranged the message, the entire radio industry suffered from goose pimples.

With that incident in mind, Censorship concluded that one open code which an enemy agent might get on the air would utilize the names of songs. The only advance preparation necessary would be selection of a station, powerful enough to be heard by his principal, which played musical numbers by request. Then the principal and the agent would work out a code in which "Deep in the Heart of Texas" could mean "Convoy leaves tomorrow"; "The Jersey Bounce" could mean "Invasion expected shortly"; and each of a hundred other songs, popular or classical, would have a meaning of its own.

All that remained would be for the agent to telephone the radio station and say: "My Aunt Martha is celebrating her eightieth birthday tomorrow. She always listens to your morning request program, and I would like to surprise her by having you dedicate to her 'Deep in the Heart of Texas.'"

In that instance, "Aunt Martha" would be the signal to the principal that a secret message was to follow.

Accordingly, Censorship asked stations to establish safeguards on all request programs. It specified that no requests should be accepted by telephone or telegraph and that mail containing requests should be held for an irregular, unspecified time before being honored on the air. In that way an agent could not be sure that his selection would be played tomorrow, the next day, or a week or ten days later.

Even Santa Claus was a Censorship casualty in this respect. Every December he had been reading lists of gifts which children desired and in some instances had been interviewing boys and girls in department stores' toylands. But those programs were out for the duration, unless the letters to Santa could be re-written or edited so as to insure purely random selection.

Broadcasters showed ingenuity in adjusting themselves to the restrictions at a time when the popularity of audience-participation shows was at its height. One small Kentucky station moved its "man on the street" program from the sidewalk to the studio, established proper safeguards in the selection of participants, and renamed it "man off the street."

Satanic Curses

All modern conflicts involve propaganda on both sides, but the recent Gulf War brought to television an ancient tradition in the Middle East, known as *hija'*—the poetry of execration. Cursing one's enemies is common in the Old Testament, and most of the cursing verses in Arabic antedate Mohammed. The latter tradition has been kept up by nomadic Bedouins still using an archaic dialect. During the Gulf War, Ehud Ya'ari and Ina Freedman monitored from Israel endless hours of cursing and counter-imprecations on neighboring Arab television stations to the accompaniment of the *rabab*, an ancient stringed instrument.

"Imagine Lord Haw-Haw spouting doggerel in Chaucerian English—to draw an imperfect parallel—and you get a sense of how these verses are received on a literal level," they wrote in the *Atlantic*, "Shriveled old geezers from half-forgotten tribes—the only people who still know how to play the instrument—have become television celebrities overnight."

The cursing contest was conducted mainly between Saudi Arabia and Iraq, with Bagdad accusing the Saudis of not being macho enough and having to hide behind foreign powers. Saudi television shot back by calling Saddam Hussein a bad neighbor, who had forgotten the favors brother Arabs provided him in the war against Iran. Then in a masterstroke, the bard managed to smear both Hussein and the eternal enemy—Israel—with the same brush, predicting a rather unpleasant retribution:

> Saddam, O Saddam,
> Of our flesh not are you.
> Claim not to be a Muslim,
> For you are truly a Jew.
> Your deeds have proved ugly,
> Your face is darkest black,
> And we will yet set fire
> To your bottom and to your back.

Shouting from the Rooftop

Seeing and listening to some of the rooftop reports from Tel Aviv and Baghdad during the recent Persian Gulf War, one could not help reflect how little some things have changed in broadcast journalism since the original rooftop reporter more than half a century before. Here is part of Edward R. Murrow's broadcast on September 21, 1940, made against the background of falling bombs and anti-aircraft fire:

"I'm standing on a roof top looking out over London . . . For reasons of national as well as personal security, I'm unable to tell you the exact location from which I'm speaking. Off to my left, far away in the distance, I can see just that faint red, angry snap of anti-aircraft bursts against the steel blue sky, but the guns are so far away that it's impossible to hear them from this location. About five minutes ago, the guns were working . . . I think probably in a minute we shall have the sound of the guns in the immediate vicinity. The lights are swinging over in this general direction now. You'll hear two explosions. There they are! Earlier this evening we heard a number of bombs go sliding and slithering across to fall several blocks away. Just overhead now the burst of anti-aircraft fire. Still the guns are not working. The searchlights now are feeling almost directly overhead. Now you'll hear two bursts a little nearer in a moment . . . There they are! That hard, stony sound."

The following, too, sounds familiar. Recollecting his London broadcasts some twenty years later, Morrow described his situation with censorship, not that different from CNN reporter Peter Arnett's experiences in Baghdad:

"I had to stand on that roof top for six nights in succession," Murrow recalled, "and make a recording each night and submit it to the Ministry of Information in order to persuade the censors that I could ad-lib without violating security. And I did it for six nights and the records were lost somewhere in the Ministry of Information.

"So I had to do it for another six nights before they would finally give me permission—after listening to the second tape of six—to stand on a roof top. So I had a lot of time up there."

How to Report

Edward Murrow instructed his wartime staff that "the reporter must never sound excited, even if bombs are falling outside. Rather the reporter should imagine that he has just returned to his hometown and that the local editor has asked him to dinner with a banker and a professor. After dinner your host asks you, 'Well, what was it like?' As you talk, the maid is passing the coffee and her boyfriend, a truck driver, is waiting for her in the kitchen and listening. You are supposed to describe things in terms that make sense to the truck driver without insulting the intelligence of the professor."

In Voce Veritas

Charles Collingwood became one of the star CBS correspondents for reporting on the North African campaigns in World War II. People thought he was cleverly trying to evade military censorship by calibrating his voice in a certain way, but he said after the war: "I honestly didn't try to avoid censorship, but sometimes I'd get so upset at the news that I guess my voice was affected."

The Voice of Doom

John Snagge became known as the Voice of London during the war, and afterward he got to cover all the major events in his capacity as Head of Presentation.

In 1954 the director-general of the BBC summoned Snagge to forbid him from making any announcements without his personal permission: "Your voice is so associated with important announcements," said the boss, "that as soon as you come on, people will assume that Winston Churchill has died."

All in a Good Cause

John Snagge's powerful and solemn voice was not versatile or suited for just any occasion. Once, when a well-known actress was sick and could not appear on *The Week's Good Cause* on behalf of a hospital charity, Snagge offered to read the script himself. "What should have been a heart-rending appeal when given by a charming woman's voice," BBC colleague Frederick Grisewood recalled in his memoirs, "was quite staggering when blurted out in John Snagge's most virile and manly tones! He became involved in the most intimate details and his embarrasment was so painful to listen to that I could hardly bear it."

Grisewood was waiting to read the nine o'clock news at an adjacent studio and was listening with intense sympathy through the headphones to Snagge's anguished description of the hospital's maternity facilities when he saw the white light in the studio flashing on. He picked up the telephone and heard one of the sound engineer's voices:

"Mr. Grisewood?"

"Yes?"

"When you next see Mr. Snagge will you tell him we've all burst into tears up here!"

Two Sides of the Coin

The humorist George Mikes was working after World War II for the Hungarian Section of the BBC's World Service. Part of his job was to beam programs behind the Iron Curtain about institutions that embodied British democratic ideals. As he went about Britain gathering material for the broadcasts, Mikes would introduce himself as a member of BBC's Hungarian Section. He would be greeted almost always with the same response:

"Good God. The BBC has a Hungarian Section? Whatever for?"

A dozen years later, the Iron Curtain was torn asunder for a few days during the Hungarian revolt of 1956. George Mikes, a famous author by then, was standing at the Austro-Hungarian border and negotiating with the freedom fighters to let him through to film the historic events in Budapest for the BBC program *Panorama*. As he introduced the team of television cameramen to his former countrymen, one of the revolutionaries, who had been listening for years on clandestine wavelengths to the BBC's Hungarian radio broadcasts, exclaimed: "Does the BBC also have a television service? Whatever for?"

Never on Sunday

When WJZ first started broadcasting in the early twenties with Vincent Lopez at the piano on Sunday nights, it violated one of the blue laws which forbade dance music on the "Sabbath." The first Sunday it happened, one of the owners of the station called the studio manager and ordered him to take Lopez off the air. In the early days, most artists performed for nothing and good talent was hard to come by. So rather than alienate Vincent Lopez, the manager simply cut off Lopez's microphone and put on a record of religious music, while the pianist banged out his lively tunes none the wiser.

A Really Silent Night

Because radio and early television went out live, many sponsors felt uncomfortable with the possibility that some performers' ad-libs would embarrass them or bring discredit to their product. Fred Friendly remembers producing the quiz show *Who Said That?* for NBC in 1949, when the sponsor kept a blacklist of proscribed names, such as Henry Morgan, Oscar Levant, Al Capp, and several politicians with controversial views. The playwright George S. Kaufman ended up on a list, not because of any political views but for his ad-lib on the Christmas program of *This Is Show Business*:

"Let's make this one program," said the wit, "on which no one sings 'Silent Night.'"

All's Well That Ends Well

One of the greatest controversies on radio was caused by an innocuous skit on the *Chase and Sanborn Hour* on December 12, 1937. Mae West was the guest with Edgar Bergen, and after a few typical remarks about Charlie McCarthy's physical endowments ("all wood and a yard long"), America's reigning sex queen went on to a skit about the Garden of Eden; she impersonated Eve opposite Don Ameche's Adam, with the dummy playing the Serpent. Arch Oboler's sketch, though a departure from the book of Genesis, certainly would be mild by today's standards. Nothing during rehearsals tipped off network executives or the sponsors about possible trouble, though Mae West held back until the actual performance on the leering innuendoes that were her specialty.

Perhaps because it was on a Sunday program, with a vast family audience, the reaction was fast and furious. Thousands protested the assault on American morals; questions were raised in Congress and a boycott threatened against Chase and Sanborn products. The company apologized through its advertising agency, J. Walter Thompson, and the president of NBC expressed regret.

"I just ran and hid," Edgar Bergen reminisced later. "The net result was that our rating went up to the highest we ever had. Our only mistake was that we were twenty-five years ahead of our time."

What the Heck

Herb Shriner, a radio comic in the forties, got into trouble with the censors at NBC with the following joke. "It's dangerous to invest in the stock market," he warned. "I once bought a hundred shares of National Underwear. The bottom fell out."

When Fred Allen, who occasionally used underwear jokes himself, heard about the censoring of this innocent double entendre, he commented: "I don't think they can bar underwear jokes unless they bar underwear."

Fred Allen ran his own war of wits against the network censors. He considered the high point of his show when he told jokes to "warm up" the audience with a ten-minute monologue and put them into a laughing mood, before the program actually began. (Nowadays, this is done by aspiring comics.) Allen would ask that if any member of the studio audience was there unintentionally, to "get the heck out of here. Heck is a place invented by the National Broadcasting Company," the comic explained. "NBC does not recognize Hell nor the Columbia network. When a bad person working for NBC dies, he goes to Heck, and when a good person dies, he goes to the Rainbow Room."

Censorship on radio and television has always been stronger than in the movies or on stage, because these media penetrate the home; broadcast and advertising executives do not want them to be tuned out or turned off. Milton Berle remembers that in the early days of television, when even words like "hell" and "damn" were forbidden, he would try to draw attention to his shackles with lines like:

"Well, I just flew in from La Guardia in a heck-acopter" and

"I'm going now to Nevada, and stop off at the Hoover Darn."

Under the Bed with Lucy

While many people who had nothing to do with Communism were smeared and driven from their jobs during the blacklist era, Lucille Ball survived the witch-hunt even though she had once joined the party, as Walter Winchell reported in his column. Phillip Morris, the tobacco company that sponsored *I Love Lucy*, fell into a panic. Just before filming the episode following Winchell's column, Desi Arnaz faced an audience composed largely of members of the international press. He explained to them that Lucille Ball's favorite grandfather, Fred Hunt, indeed was an old-time Socialist who had registered himself and Lucy in support of a Communist candidate for mayor of Los Angeles. Grandpa had been a real character, and Desi told as many stories about his lovable eccentricities as possible. According to columnist James Bacon, he wound up his brief speech with a flourish:

"The only thing red about Lucy is her hair—and even that's not real."

The audience laughed, the press dismissed the case, and Phillip Morris officials could breathe again, as *I Love Lucy* continued to score an 80 share audience for its sponsor.

Robin Hood

During the McCarthy era, many of the blacklisted Hollywood writers and actors left the West Coast. Some went to Europe, while others tried to rebuild their shattered careers in New York.

Fortunately, television drama was just beginning in the early fifties, and a few of the writers found employment on such series as *The Adventures of Robin Hood*, the theme of which fitted their own beliefs about social justice. In fact, some of the dialogue takes on ironic meaning when we know that Ring Lardner, Jr., or Waldo Salt had written those lines.

At the time, of course, the proscribed screenwriters could not work under their own names and were forced to employ pseudonyms. The name of Waldo Salt's wife, M. L. Davenport, was used not only by her husband (as Mel Davenport) but also by others among their friends.

"Yes, Davenport spent an awful lot of time at his typewriter," Mary Davenport reminisced in the Oscar-winning documentary about her late husband.

Thank You, Doctor

Most major phone-in shows employ a screener to keep the lunatic fringe off the air. Bill Miller was once chatting with a group of surgeons on his late-night radio program *Contact*, beamed from Schenectady, New York, when a man called, who knew how to get past the screener. He began pleasantly enough by thanking the guests for giving him a lot of work.

"What line of work are you in?" asked one of the doctors.

"I work with an undertaker," the caller replied. "And you sons-of-bitches are responsible for everybody we put in the cemetery!"

Undeleted Expletives

For those who get on the air past the screening system, there is the tape delay button which the talk-show host can press, allowing an engineer to bleep over the offending word. Hilly Rose tells a number of stories in his book *That's Not What I Called About*, when the machine worked less than perfectly, deleting everything except the expletive.

Paul Benzaquin was suspended for a week at WBZ in Boston, when he failed to censor a playwright on his show. He tried, but instead of deletion audiences heard "horseshit" repeated several times on a loop.

Rose himself recalls once having to eject from his studio a visiting author who felt cornered by his probing questions and blurted out:

"You and your fucking facts!"

Emphatically No

In the early days of television, a code was promulgated regarding interviews with bosomy film and stage stars popular at the time. "No emphasis on anatomical details," was one instruction to the cameramen.

Ben Gross, broadcast critic of the New York *Daily News*, asked Dagmar, the Danish sexpot, what she thought of this new rule.

"Honey," she replied, "I don't need any emphasis."

I'm Glad You Asked

David Frost asked Joan Crawford in a television interview whom she considered during her long career the sexiest man in the movies.

"Clark Gable," the star responded without hesitation.

Frost asked why, and was startled—along with the censor—when she replied in one word: "Balls!"

Crime and Punishment

In a 1965 debate on the BBC about censorship, critic Kenneth Tynan argued with novelist Mary McCarthy about allowing intercourse to be depicted on the stage of the National Theatre: "I doubt if there are very many rational people in this world," he observed, "to whom the word 'fuck' is particularly diabolical or revolting or totally forbidden." This was the first time that this particular four-letter word had been uttered on the British airwaves, and the resulting controversy eclipsed even the memorable occasion back in 1914, when Mrs. Patrick Campbell first said "bloody" in the role of Eliza Doolittle.

Headlines in the newspapers and motions in Parliament demanded immediate action against both Tynan and the BBC; the clergy and schoolmasters were up in arms. Harold Wilson's government withstood the pressure to prosecute the errant critic, and the prime minister jokingly promised to refrain from the use of four-letter words in his television addresses. Mrs. Mary Whitehouse, who had founded the Clean-Up TV campaign, wrote to the Queen to use her influence to remove filth from television. The enraged housewife also suggested that Kenneth Tynan needed his bottom smacked. And indeed, the late critic—who was yet to conceive the long-running pornographic hit *Oh! Calcutta!*—might have enjoyed such kinky punishment.

On the Alert

Television censorship is usually self-inflicted and sometimes takes bizarre forms. In his book, Eyes As Big As Cantaloupes, *popular TV columnist Don Freeman recounts one example.*

In the early 1970s, NBC had in readiness the notion for a Saturday morning children's show called *Land of the Lost*. In it, Rick Marshall, a forest ranger, and his teenage son and daughter, Will and Holly, are floating down a river on a raft. Then an earthquake hits and they are transported by a treacherous waterfall into a prehistoric world lost somewhere in the vast corridors of time.

There they are confronted with strange reptiles, dinosaurs, flying animals and the native monkey-men, known as the Pakuni tribe—and, as every right-thinking viewer soon learned, the tribe speaks flawless Paku.

Paku is a real, made-up television language, a linguistic first and, possibly, a last. When the show was conceived high in the corporate headquarters of the National Broadcasting Co., a vice president named Joe Taritero reasoned: "What I don't want in this show is the monkey-men going around saying 'umgawa.' I want their sounds to be a real language. Maybe *a whole new language!*"

Taritero, seeking a whole new language, was referred to Dr. Victoria Fromkin, a renowned scholar and head of the Department of Linguistics at UCLA.

"Need-um new language," Taritero said, more or less.

"You got it," said Dr. Fromkin, not in those exact words.

The creation of Paku emerged as a fascinating and scholarly project. This is by no means a form of pig latin. Paku is—to use a phrase not yet found in Paku—the real McCoy. Although her colleagues were amused by Dr. Fromkin's association with children's television, known to readers of *Variety* as "kid-vid," she went at her work with dedicated fervor. She devised the Paku language according to the strict international rules of syntax. Plurals, for example, were formed not by adding "s," as in English, but by adding "ni" (pronounced knee). Thus, the Paku word for child is "abu"—for children it's "abuni."

In the very first *Land of the Lost* script that she received, there were indications that two Pakuni were swearing. Accordingly, she made up some Paku obscenities. NBC, naturally, censored them out.

Incensed Censor

Saturday Night Live had problems with censorship from the beginning. Even before the late-night satirical show went on the air in 1975, producer Lorne Michaels received instructions from NBC lawyers about what would be allowed or forbidden.

"I've understood everything you said," Michaels told one of them. "But if the show is a failure, nothing you've said really matters. And if the show is a success, then nothing you've said really matters."

During its many years on the air, *Saturday Night Live* would come to test the limits of Broadcast Standards, as censorship is generally euphemized in the electronic media. The censors sometimes simply did not get some of the double entendres in the current argot of New York, or they took at face value promises that the actors would adhere to the text as written. In the end, the temptation posed by a live microphone often proved irresistible. Burt Reynolds, in a parody of a sleazy commercial for a photo album, urged ladies to get their rocks off by buying "The Burt Book." And the censor, hearing in the control room an off-color joke improvised by Dan Ackroyd, had so far forgot himself and his profession that he leapt to his feet, yelling:

"I've been screwed!"

Stop and Think

Mike Wallace had his first experience with being sued long before *60 Minutes.* *
During a 1957 live interview with Mickey Cohen on ABC's *Nightbeat*, the mobster vented his unvarnished opinions about Los Angeles police chief William Parker and his deputy, Captain James Hamilton. Both Cohen and Wallace were sued for libel.

After settling out of court, the network insisted that all of Wallace's interviews must be attended by a lawyer, who sat with him off-camera. Wallace frequently

* General Westmoreland took him to court in 1984–85 for a Vietnam documentary.

recalls how every time his line of questioning wandered off into litigious territory, the lawyer held up a cue card, reading "CAREFUL!" or "STOP NOW!"

A Touch of the Poet

Until the introduction of penicillin, syphilis was one of the most dreaded diseases. The problem was complicated by censorship which did not allow even the word to be mentioned on radio or newsreels.

Finally, under pressure from opinion makers and newspaper editorials, the prohibition was removed. The first time the subject came up, an announcer on *The March of Time* explained: "This terrible disease which we must all fight is said to get its name from that famous Greek poet, Syphilis."

CHAPTER 6

Public Affairs

Fireside Chat

Franklin Delano Roosevelt's famous fireside chats began in March 1933, just after he took office. The chief executive announced his intention of addressing the nation, and when the networks sent technicians to set up microphones, it was FDR who suggested the location, in front of the fireplace in the Diplomatic Reception Room of the White House.

"I can sit down in cozy informal surroundings," FDR is quoted by veteran CBS newsman David Schoenbrun, "and just have a chat with my friends and neighbors, instead of a windy speech."

According to the same source, it was Harry C. Butcher, manager of the CBS bureau in Washington, who christened these presidential addresses, never rivaled for intimacy since, as "fireside chats."

Peter Grant, who announced Red Skelton's radio show, was almost taken off the air because people thought he was trying to mimic Franklin D. Roosevelt's voice. In fact, the resemblance was natural. Skelton defended him: "You can't take a man's livelihood away from him just because he sounds like the president." Grant made a conscious effort to change his delivery, and stayed on the air.

Payback

When America entered World War II, Arthur Godfrey was eager to serve, but the Navy refused him a commission because of some old hip and knee injuries. Somehow, the popular broadcaster got word to the White House, where Eleanor Roosevelt went to plead his case to her husband, who had spent much of his life in a wheelchair. FDR placed a call to the Navy—which he had commanded once—and said:

"Can he walk? Give him a commission, then. I can't walk and I'm the commander-in-chief!"

The intervention may have given that extra emotional wallop to Arthur Godfrey's voice, when a few years later he covered FDR's funeral on CBS radio.

Entity

Along with Eleanor Roosevelt and Gabriel Heatter, H. V. Kaltenborn was one of the most mimicked voices on radio. During a Kansas City dinner Harry S Truman also did a parody of one of Kaltenborn's 1948 election broadcasts, the gist of which was that Truman could not possibly win.

"I don't mind being imitated," the commentator said. "After all, they don't make fun of non-entities."

The S.O.B.

Drew Pearson was one of the most influential columnists and broadcasters in the forties and fifties. He frequently accused members of the Truman administration of accepting patronage from lobbyists.

"I want it distinctly understood," the plain-speaking president announced at a public banquet, "that any S.O.B. who thinks he can cause any of these people to be discharged by me, by some smart aleck over the air or in the paper, has another think coming."

Pearson explained to his listeners that he understood the letters S.O.B. to mean "Servants of Brotherhood."

"Some people should learn the alphabet," Harry Truman fumed to an aide. "S.O.B. is as simple as ABC."

The epithet stuck, and Pearson became so widely known that letters addressed to "The S.O.B—Washington, D.C." were delivered to the journalist's home. He even toyed with the idea of having it engraved on his tombstone as his sole epitaph.

A Save by Nixon

Another of Drew Pearson's targets was Senator Joseph McCarthy, who responded by organizing a boycott of Adams Hats, the company that sponsored Pearson's broadcasts. Their antagonism came to blows after a dinner they attended at Washington's Sulgrave Club. According to Pearson, McCarthy came up to him in the cloakroom, kicked him in the groin, and slapped him in the ear. The junior senator from Wisconsin only owned up to the slap, which caused a young congressman named Richard M. Nixon to step between them. "I never saw a man slapped so hard," Nixon was quoted as saying later. "If I hadn't pulled McCarthy away he might have killed Pearson."

Giving 'Em Hell

Harry Truman went home to Independence, Missouri, just before the 1948 presidential elections to await the returns. The press, having forecast his defeat to Governor Dewey, was anxious to register every moment of his humiliation, and pursued him everywhere.

Lewis W. Schollenberger, who had covered presidential broadcasts for CBS since the early forties, recalled that the press plane landed at Kansas City at the same time that a minor accident occurred at the airport. In the ensuing confusion the press corps lost sight of its quarry; Truman had already been driven off toward Independence.

The press rushed to waiting cars, and they even managed to commandeer some of the local police to clear traffic for them. With flashing lights and blaring sirens, the corps arrived in front of the modest Truman home, but there was no sign of the president. After a few minutes, his car did pull up, and as Truman got out, one of the journalists asked:

"What happened, Mr. President? Where were you?"

"Oh, we were stopped and had to pull over," Truman grinned. "Seems there were some very important people going through town."

Cue for Action

General Eisenhower, appointed by President Truman to command Allied Forces in Europe, made a televised speech from his office at the Pentagon. Since Eisenhower hated microphones, CBS announcer Lew Schollenberger hid one in a telephone, and another behind a large globe.

Cue cards were just beginning to come in, but there was no rack to hold them, so Schollenberger commandeered two Army colonels instead. They stood back to back, one facing the general, holding up a card, and the other in reserve, facing

Schollenberger, until he gave the signal to execute an about-face, which provided the next cue card.

In the fifties, the networks would cut off even the president during a paid political broadcast, if he ran over. It happened during a rally when Eisenhower ignored Schollenberger's frantic gestures that he was running into unpaid overtime.

A few days later, during a presidential address, Ike looked off-camera at Lew Schollenberger and explained:

"I understand the other night there was a young man here signaling to me that I was behind schedule. I never saw him."

Adlai Who?

Somewhat reluctantly, General Eisenhower adopted television as a weapon to fight the 1952 presidential campaign. It was the first time that slick political commercials showed a candidate dealing with major issues with smiles and vague sound bites.

The Democratic campaign, fearing that these ads might prove effective, suggested that its candidate, Adlai Stevenson, should counterattack with the same tactics.

"This is the worst thing I've ever heard of," Stevenson replied to a political adviser. "How can you talk seriously about issues with one-minute spots!"

First Things First

During the presidential election campaign of 1952, the candidates appeared on television for a total of two hours and ten minutes. Meanwhile, during the same period, Betty Furness consumed four and a half hours to sell household appliances.

Ike Liked Him

The Washington television and radio press corps invited Phil Silvers to emcee their 1954 annual satirical show presented to entertain their natural enemies. This one was attended by President Eisenhower, Vice President Nixon, most of the Cabinet, and many members of Congress. Silvers was edgy because of the elaborate security arrangements, heightened by the general atmosphere of the McCarthy era. The comic surveyed the assembled lawmakers and administrators, waited for fifteen seconds, and then turned to the chief executive:

"My God, who's minding the store?"

Dull As Dulles

At the beginning of her career, Carol Burnett became famous for a song, "I made a fool of myself over John Foster Dulles." She had meant it as a satire about the adulation young girls were giving Elvis Presley, but the song became major news after she sang it for Jack Paar and then Ed Sullivan:

> I made a fool of myself over John Foster Dulles,
> I made an ass of myself over John Foster Dulles
> The first time I saw him 'twas at the UN . . .

The dour and unlovely-looking secretary of state apparently liked the song enough to request a repeat performance on *The Tonight Show*. When he was questioned on the program *Meet the Press* whether there was anything going on between him and the bright young comedienne, he replied gravely:
 "I make it a policy never to discuss matters of the heart in public."

Close Calls

During the closely fought general election of 1964, Harold Wilson was worried that Labour voters might stay away from the polling booths because of the popular BBC program *Steptoe and Son*. When Sir Hugh Greene postponed the broadcast until 9 o'clock, Wilson called to thank him, saying that the decision might swing twelve seats. Wilson and Labour won by just four seats.

John F. Kennedy's victory over Richard Nixon had the smallest margin in American history, and Kennedy freely attributed it to television. "We wouldn't have had a prayer without that gadget," he said after he knew the final results.
 Many believe that Nixon lost the election during the first televised debate with his opponent because of the famous five o'clock shadow. But his unshaven look was only a small part of the problem. Kennedy came well rehearsed, tanned from campaigning in California, needing no makeup. Nixon, who began his journey to the White House with his famous "Checkers" speech eight years earlier, was constantly photographed as vice president, and felt overly confident about the medium. He hardly had any preparation. Also, having to stand behind a lectern, rather than sitting in an armchair, helped the athletic and youthful Kennedy look magisterial, while Nixon, suffering from a bad leg, kept continually shifting the weight on his feet.
 Afterward, the Nixon camp was crying foul, claiming that the camera cut away from Kennedy to take reaction shots of Nixon only when the latter was mopping his brow. They pointed out that Kennedy was in a dark suit which showed up well against the light background, while CBS had told Nixon to wear a light-colored

suit. The vice president's people tried to have the flats repainted, but still Nixon washed out into the decor.

Interestingly, a poll taken after the debate showed that those who heard it on radio thought that Nixon had won, whereas the 75 million who saw it on television gave a definite edge to Kennedy.

Although he seemed always nervous and ill at ease on television, Richard Nixon apparently believed that he looked good on the small screen, which had chronicled so many high and low points in his career. As president, he got into the habit of wearing pancake makeup quite regularly, just in case he might be captured on television.

Not All the Way

Lyndon Johnson was sensitive about his jug ears, and he never showed his profile to the camera. Johnson owed much to television: much of his personal fortune came from his wife Lady Bird's shares in a station in Texas. But he also blamed the medium for losing what came to be known as "the living room war" in Vietnam.

LBJ would often call broadcast executives or reporters to give his instant reaction to a particular newscast or documentary. After he heard one of Morley Safer's reports from Vietnam, Johnson called Frank Stanton, president of CBS, early the following morning.

"Frank, are you trying to fuck me?" asked the commander-in-chief. "Yesterday your boys shat on the American flag."

On one occasion, when Johnson was in New York having just settled a railroad strike, he impulsively barged into CBS to make the announcement in person. Called to account by his wife for his impulsive behavior, the president told Lady Bird:

"I wanted to see the look on Walter Cronkite's face when I walked in the studio."

(Some have thought that Johnson decided not to run for a second term after Walter Cronkite broadcast a news special about Vietnam, in which America's most trusted man concluded that the United States should get out of Southeast Asia. Johnson told his press secretary that if he had lost Walter Cronkite, he had lost Middle America.)

Back on the Plantation

Bill Moyers, the ubiquitous presence on American public television, was a close advisor and then press secretary to Lyndon Johnson. Before some presidential press conferences, Moyers resorted to the practice of planting certain important questions with reliable journalists. But perhaps because he had trained to be a Baptist preacher, the underhandedness bothered Moyers. He once opened a press conference with:

"I'll take the planted questions first."

Imagine, If You Will

Mike Wallace, at the time Don Hewitt approached him to become one of the reporters on *60 Minutes*, was considering another job—that of press spokesman for President Richard Nixon. Hewitt talked him out of it by declaring that "a press spokesman is a nobody trying to be somebody."

Diane Sawyer, on the other hand, put in eight years with Nixon as a writer and researcher, during and after his presidency. It was a great deal to live down and atone for. When she became the first female reporter on *60 Minutes*, by then the top-rated public affairs program in America, she tried to emulate Mike Wallace in grilling people. In interviewing George Bush, then vice president, Sawyer cited the fact that he had been called a wimp.

"Are you a wimp?" she asked.

On another program, Diane Sawyer visited with John Connally, the former Texas governor, after he had lost his fortune in the oil slump of the early eighties. Talking to him about the fact that even his house and personal possessions had to be auctioned off, Sawyer asked the broke and broken old man how it could have happened.

"Were you stupid?" she prompted helpfully.

Ex Cathedra

At the end of a summit meeting in the Bahamas between President Kennedy and Prime Minister Harold Macmillan, White House press secretary Pierre Salinger briefed waiting reporters about the final communiqué:

"The statement will be read by both the president and the prime minister personally."

"Both of them will read the same statement?" asked one of the newsmen.

"They will both be together," Salinger replied with a hint of irritation. "They are not going to sing *a capella.*"

Mac Attack

Harold Macmillan was the first British prime minister who became conscious of having to project an image on television. He was not entirely comfortable about the medium, which he compared to "playing lawn tennis and there isn't anybody to hit the ball back from the other side of the net." In fact, Macmillan learned to return the ball quite well himself. In a famous episode, during his 1959 visit to Moscow, Macmillan was asked by Robin Day of Independent Television News about the date of a long expected general election. Macmillan assumed that Day saw his Soviet tour as a mere election ploy, a view that was widely held at the time, and lashed back at him:

"That is a question which, if I may make a plagiarism, is made by the wrong man in the wrong place."*

Though Harold Macmillan was sensitive to criticism, he also stood up for freedom of expression. British political satire was enjoying a renaissance in the early 1960s, but many people in the establishment were shocked when the BBC joined the trend with the landmark program *That Was the Week That Was.* Conservatives mounted pressure to force it off the air, but Macmillan, himself the chief butt of the political sketches, intervened. Reporters at one point asked the postmaster general, Reginald Bevins, whether he would do something about *TW3*, as the program was called, and he told them firmly:

"Yes, I will." The very next morning, the tersest memo arrived from the prime minister, which just said:

"Oh, no, you won't."

Pardon Me

Chevy Chase became the first star on the late-night satirical review *Saturday Night Live* with his newscasts (which began, "I'm Chevy Chase, and you're not") and his pratfalls in imitation of a clumsy Gerald Ford. The president, with his good sense of humor, readily agreed to address the annual dinner of the Radio and Television Correspondents Association in 1976, even after he was told that Chevy Chase would provide the evening's entertainment.

* General Omar Bradley had commented, when General Douglas MacArthur suggested that the Korean War should be expanded into China: "The wrong war, at the wrong place, at the wrong time, and with the wrong enemy."

Determined not to be upstaged, Ford had secretly rehearsed a routine of his own. As he stood up to speak, he deliberately pulled the whole tablecloth with him. Once at the podium, he had a sheaf of blank pages that he scattered by seeming accident all over the floor. Finally, the president stole the show by pointing at his impersonator with his opening comment:

"I'm Gerald Ford and you're not." It got a big laugh. "Mr. Chevy Chase," he went on, "you are a very, very funny suburb."

Later that year, Gerald Ford's press secretary, Ron Nessen, who had once worked for NBC News, hosted *Saturday Night* on the principle that he might help his boss's re-election campaign. After all, even Richard Nixon appeared in 1968 on *Laugh-In* and left America speechless with his one line: "Sock it to me!" But the particular episode that Nessen hosted (and which included taped inserts with Gerald Ford) was generally nasty, tasteless, and had the opposite effect from what Nessen had intended. *Washington Post* columnist Bill Gould quoted one reader questioning the president's judgment in allowing his spokesman to appear:

"I don't see how I can vote for a man who could be so dumb."

Movin' In

When President Gerald Ford found out that he was staying at the same hotel as Claude Akins, star of the television show *Movin' On*, he invited the actor to his suite. He told him he was a fan, but that his main motive was so that he could brag about his meeting to his wife Betty.

Betty Ford's comment during a White House interview on *60 Minutes* that she wouldn't be surprised if her daughter Susan had had an affair caused a national turmoil almost matching that of her husband's decision to pardon Richard Nixon. Though a poll showed that most women supported her position, the First Lady drew fire from the right wing of the Republican party. At a press briefing, White House spokesman Ron Nessen was asked how Gerald Ford reacted to the interview. This was the official answer:

"The president has long ceased to be perturbed or surprised by his wife's remarks."

How to Get a Good Night's Sleep

In the mid-seventies, Gordon Manning went from CBS to NBC, where he engineered some spectacular political scoops, especially involving the Chinese and the Russians. On the domestic front he laid a few eggs, such as the spectacular failure of producing President Ford's memoirs on television.

"I think Gerald Ford went to sleep on the air before I did," John Chancellor reminisced, "but I went to sleep shortly after that."

Bogeyman

Sam Donaldson made his national reputation with his fearless and strident questions during the Carter and Reagan administrations. Since Ronald Reagan is somewhat deaf, and his preferred mode of talking to the press was under the whirling blades of the presidential helicopter, some might say that Donaldson's noisy approach was necessary.

In the final phase of the 1984 election campaign, on Halloween, the president was walking to the back of his cabin on Air Force One when he was suddenly confronted by someone with a ghoulish mask. Reagan peered at the masked man and quipped:

"Phew, I thought for a moment you were Sam Donaldson."

While Reagan Was Awake

Ronald Reagan began his career in the early 1930s as a sports announcer on radio station WOC in Davenport, Iowa. The letters stood for World of Chiropractic, since the pioneering station had been founded by Colonel B. J. Palmer, who ran a chiropractic school in the same building. The colonel also owned WHO in Des Moines, and it was from this station that young Reagan announced all kinds of sporting events for NBC, including six hundred big-league baseball games. He also interviewed many sports personalities, such as Max Baer, who accidentally knocked down a mailroom clerk in the studio while demonstrating a punch.

At the same station the future president endured one of the most embarrassing moments of his professional career. He had to interview the famous evangelist, Aimee Semple McPherson, who came to publicize a revival meeting in Des Moines. "Suddenly I heard her say goodnight to the audience," Reagan recalled, "and I was sitting there with four minutes to fill." The young broadcaster desperately signaled to the engineer to put on a record and announced that the broadcast would conclude with a brief interlude. But instead of the solemn, sacred music Reagan anticipated, the engineer reached for the nearest stack of records, and Miss McPherson's uplifting program ended with the Mills Brothers singing "Minnie the Moocher's Wedding Day."

Ronald Reagan's political success was greatly helped along by his opponents who constantly dismissed him as a mere actor. When he was running to become governor of California, the campaign committee for the veteran Democratic incumbent, Pat Brown, ran a nasty television spot in which the governor was telling an elementary classroom: "I'm running against an actor, and you know who shot Lincoln, dontcha?"

Reagan won that race, and at his inauguration, just past midnight on January 2, 1967, spotting in the crowd George Murphy, another actor-turned-politician, he quipped: "Well, George, here we are on the late show again!"

Flacks claimed at the time that the midnight swearing-in ceremony symbolized Governor Reagan's eagerness to get down to work, though twenty years later many believe that the auspicious occasion had been arranged after consultation with Nancy Reagan's astrologer. At any rate, by the time he got to the White House, Reagan showed little inclination for work. Lou Cannon, in his recent book on the Reagan presidency, recounts an incident just before the economic summit held at Williamsburg in 1983.

James Baker, then White House chief of staff, had prepared a large briefing book which he had left for Reagan to study before the conference. He was shocked to find it the next morning exactly where he had left it. He asked his boss why he had not even opened the book.

"Well, Jim," the president replied in all seriousness, *"The Sound of Music* was on last night."

The Late Late Show

Hubert Humphrey, when he was vice president, told the annual dinner of the Gridiron Club that he enjoyed chairing the Senate's sessions:

"Where else would I be able to get an occasional glimpse of Senator George Murphy without makeup? They tell me that when George Murphy entered the Senate chambers for the first time, he walked up to a page and said, 'Where's my dressing room?'

"In a way it's nice to have George Murphy in the Senate. During a lull in the proceedings he lets me read his *Variety.*

"It's too bad they voted down Mr. Murphy's first proposal. He suggested the Senate inaugurate late, late sessions!"

Peter Ustinov remarked on the BBC program *Any Questions?* (in 1968): "Two members of my profession who are not urgently needed by my profession, Mr. Ronald Reagan and Mr. George Murphy, entered politics, and they've done extremely well. Since there has been no reciprocal tendency in the other direction, it suggests to me that our job is still more difficult than their new one."

Programmer for a Day

In another of his attempts to reaffirm family values, President George Bush recently told a group of religious broadcasters:

"We need a nation closer to *The Waltons* than to *The Simpsons*."

But Bart Simpson, the wisecracking teenager in the irreverent cartoon program, had an answer for the president:

"We're just like the Waltons. We're praying for the end of a depression, too."

Hasta la Vista

At the time of the Cuban revolution in 1960, no fewer than 62 American television series were running in Havana. As these were cut off by the U.S. embargo, Cuban TV was forced to fall back on prints of old movies on the island. Many of these were also American films, and Ginger Rogers was widely rumored to be Fidel Castro's favorite star, other than Fidel himself.

No totalitarian leader has dominated the airwaves so personally and exclusively as Castro, who likes to give rambling speeches lasting for several hours. He was so addicted to appearing on television that the plotters in one coup, in April 1964, reportedly tried to electrocute him through the microphone. To the Cuban people, used to brownouts and every other kind of shortages, Fidel's appearances have been a mixed blessing.

"The only time we can be sure of electricity," said one Havana housewife, "is when Castro speaks. They want us to see him on television."

CHAPTER 7

Views and Interviews

Return to Sender

Margaret Truman was booked to sing and to be interviewed on *The Big Show*, but prior to her appearance, the president's daughter provided a long list of topics that she refused to discuss. Understandably, she wanted to avoid politics, or any mention of her father at the White House. One skit she vetoed involved Phil Silvers, who was supposed to help her break into show business through an audition at the Copacabana. Part of the humor came from the fact that Silvers failed to recognize the First Daughter. Finally the comedian asked for her address, and Margaret Truman was to reply: "1600 Pennsylvania Avenue." Then Silvers would have asked: "Is that an apartment house or hotel?"

After Truman threw out the whole skit, writer Goodman Ace asked why she had objected.

"I don't want any cranks to know my address," she replied.

Dumbstruck

One of Robert MacNeil's early assignments working for the local CBC television station covering the Ottawa and Hull area was reporting live from the Ottawa

Valley Exhibition, an annual agricultural fair. He stood between the director of the event and his assistant, and began:

"Before seeing the sights of this year's 'Ex,' we're fortunate to have with us the man who has made this Exhibition one of the leading events of this region for the past twenty-seven years, the director, Mr. . . . I'm terribly sorry but I seem to have forgotten your name."

The director kindly helped him out and the young reporter began his interview, when he realized that he would soon have to switch and talk to the assistant, whose name he had also forgotten. Trying his hardest to remember, while prolonging the first interview, MacNeil was finally forced by the frantic urging of his floor manager to turn to the man at his other side.

"You're not going to believe this," the newsman tried to smile, "but I've forgotten you name, too!" The man believed him, MacNeil recalled in his book of memoirs, *The Right Place at the Right Time*, "and didn't look too happy about it."

Robert MacNeil, of course, went on to become one of America's most accomplished interviewers as the co-host of the daily PBS news program, the *MacNeil-Lehrer NewsHour*.

Just Wunnerful

What makes a good interviewer is the ability to listen, and the opposite is also true. According to Jack Paar, television interviewers are always desperately trying to think up their next question, and if they become aware that their guest has stopped talking, they would say automatically, "That's wunnerful." Paar recalled in his book, *My Saber Is Bent*, one program during which a patient in an iron lung and an attendant nurse were explaining how the life-saving machine worked, ending with the caveat that if the electricity went off, alas the patient would die.

"That's wunnerful," said the host.

Another of Paar's favorite stories on the subject was Cornelius Vanderbilt, Jr., recounting to a Hollywood interviewer some of his journalistic adventures during World War II.

"I was covering the fighting on the Russian front and one day I was captured by the Russians. I was thrown into an armored car and driven all night to an unknown destination. When I was dragged out of the car I was stunned to see we were at the Kremlin. My captors hauled me into that forbidding bastion, down a long gloomy corridor, and finally hurled me to the floor. Looking up, I saw Stalin glowering down at me!"

At this cliffhanger, Vanderbilt paused for breath and dramatic effect.

"I see," said the interviewer. "Do you have any hobbies?"

Not His Funeral

Bob Elson, longtime sports announcer on WGN in Chicago, was just such an absent-minded interviewer. He had a program consisting of going down to the train station and talking to celebrities who pulled into town on the Twentieth Century Limited on their journey from either coast. Once he encountered Danny Kaye and asked his standard question:

"What are you doing in town?" Danny Kaye replied that he had come for his mother's funeral.

"Are you going to be seeing any shows during your visit here?" Elson went on. The comedian also went on, but about the loss of his mother, and how much she had meant to him. Not listening, Elson asked his next standard question:

"Will you be enjoying any of our local restaurants while you are here?"

The Wright Stuff

In a live interview with famed architect Frank Lloyd Wright, Mike Wallace was chain-smoking on the air, something that was allowed in the unenlightened fifties. In fact, the interview program was sponsored by the Philip Morris Tobacco Company.

The smoke was bothering Wright, who refused to answer questions, and instead kept talking about his interviewer's bad habit.

"Is this something you feel like you should apologize for?" Wright asked Mike Wallace.

"No," the reporter squirmed uncomfortably on live television, "but can I offer you one?"

"I wouldn't know what to do with it," the old curmudgeon replied with infinite contempt.

Frank Lloyd Wright was a notoriously difficult guest. He had walked off the set in the middle of an interview with Alistair Cooke on *Omnibus*, and again with Charles Collingwood on a program prophetically called *Adventure*.

In spite of this, Garry Moore was determined to have the eighty-seven-year-old legend on his show in 1956. After an exploratory discussion in Wright's suite at the Plaza Hotel—Moore was cautioned not to smoke or mention that one of his earlier programs had been sponsored by a cigarette company—the architect came on the show. "The first thing he did," Herb Sanford recalled, "after his entrance and greeting, was to speak contemptuously of television commercials. I thought he was testing to see what sort of response he'd get from Garry. But I also thought—here goes the feature!"

In fact, the two of them got on very well, and Wright came to do the show dramatically dressed in a cape and maroon muffler, wielding a silver-headed cane. He was in total command of the interview, and apart from criticizing the design of the set, behaved himself. He was such a hit that Wright himself asked to come back for an encore visit.

Enjoying his first fling with television, Frank Lloyd Wright accepted other invitations. Appearing as a guest on *What's My Line?* he was asked by Faye Emerson:

"Mr. Wright, what do you think of New York City?" Without a beat, the great architect replied:

"Miss Emerson, I never think of New York City."

Small Expectations

In a typical episode of the media treating the media as news, Mike Wallace interviewed David Frost just prior to the latter's series of interviews with Richard Nixon in the spring of 1977.

"If you feel he is stonewalling or lying," Wallace asked, "what will you do?"

"I shall say so again and again," Frost replied. "I should not want to give the impression that he is lying."

"Do you expect him to lie?"

"No, what I expect is a cascade of candor."

"A cascade of candor—from Richard Nixon?" Wallace jumped on Frost. "Is that what you expect?"

"No," Frost shrugged, "it's just a phrase I thought would appeal to you."

She Wanted to Be Alone

Chris Wallace was starting an interview on *The Today Show* with Ann Haldeman, the daughter of Richard Nixon's former chief of staff, who was so nervous that she froze. She remained silent, looking distressed, as Wallace tried a number of different approaches. Finally, the interviewer leaned forward and patted his guest in an attempt to reassure her.

"It's okay," Wallace said. "It's just you and me sitting here, having a conversation."

Even though several million people were eavesdropping, Haldeman relaxed and breezed through the rest of the interview.

Start Talking

Ira Fistell, the popular late-night talk-show host on KABC radio in Los Angeles, got into broadcasting when he was teaching at the University of Wisconsin in Madison, working on his Ph.D. in history. He and his friend Danny got a break, and their baptism of fire, when a local station changed its format to Talk Radio.

Came first day on the job: it was the day also that the station was switching to the new format, and management was swarming all over the place, though it was the weekend. My friend Danny went on the air first. I went down to the station to give him moral support. He was very nervous and kept asking, "What if nobody calls?"

After the news and weather, Danny started on a topic and announced that the switchboard was open. No calls. He reads the weather again. And again. And again. After forty minutes, still no one called. Danny couldn't think of anything more to say. Everyone got into a huddle—program director, station manager, and various consultants—but none of them could figure out why nobody had been calling or why the new format wasn't working. At around 11 P.M., the fifty-dollar-a-week switchboard operator came waltzing in, sat down at the PBX board, flipped a couple of switches, and suddenly the studio is flooded with calls. No one had remembered to tell her about the change in format, so she had set the switchboard, as usual, to direct all incoming calls to the newsroom. Fortunately, she had been listening to the station at home and figured out what had happened.

Danny had been through the ringer that night. He took a few calls, but was exhausted. About 11:30, the general manager said to me, "Why don't you take over at midnight?" I said, "Fine."

Well, I sat down in front of the microphone and everything clicked. I have never, ever, been nervous for one second on the air. It was like I was born to do this, from the very moment I started broadcasting.

The next morning I got a call to come down to the station. I thought, "Are they going to fire us?" The general manager said:

"Ira, your friend quit this morning. How would you like to do the whole evening show, five nights a week?"

And that's how I got into radio in 1968. It turned out to be the best possible time in the history of radio to start a talk show in a town-and-gown community like Madison. It proved to be a momentous week. Not only did Lyndon Johnson give up running on Sunday night; on Tuesday night Eugene McCarthy won the Wisconsin primary, which was really a referendum against the Vietnam war. And on Thursday, Martin Luther King, Jr., was assassinated. Of all the weeks to begin a talk show, that was it.

Talk Radio

English-born Michael Jackson, the long-reigning king of KABC's Talk Radio in Los Angeles, has impeccable manners. He never loses his cool, no matter how tempted by some of his guests. He was interviewing Charles Colson once—the man who claimed that he would run over his own grandmother to re-elect Richard Nixon. Colson was on the program promoting his book about how he had been "born again" in jail, and sensed some skepticism on Jackson's part. Immediately after the program went off the air, the former Watergate convict took the broadcaster's hand and looking deeply into his eyes, said:

"I don't think, Michael, you believe I found God." Jackson looked straight into Colson's soul and replied:

"Oh yes, I do, sir. I only wish you had found Him earlier."

Reminiscing with Hubert Humphrey, KABC talk-show host Michael Jackson asked:

"Sir, you were Lyndon Johnson's vice president. As next in line of succession, didn't you ever get the urge to break his neck?"

"Let's just say leg, Michael," Humphrey replied.

Falling Apart

As everybody waited for Lyndon Johnson to announce his running mate for the 1964 presidential election campaign, a local television reporter cornered Hubert H. Humphrey on the convention floor.

"Senator, have you heard anything? Are you nervous? How are you bearing up?"

"Obviously a lot better than you are," HHH replied.

The Iron Lady

Margaret Thatcher, whether questioned in the House of Commons or on television, usually manages to say exactly what she wants. One prominent British journalist, Peter Jay, summed up his experience: "Interviewing Margaret Thatcher is very like interviewing a telephone answering-machine: you pose a question, you get an answer and you start to make a response and you find it's just going on and on and on."

Trying to draw out Mrs. Thatcher on the subject of retirement after being prime minister for a decade, David Frost tried another approach.

"Do you think," he asked during a 1989 television interview, "there will ever be another male prime minister?"

"I think," replied the indomitable Iron Lady, "one day male prime ministers will come back into fashion."

Japan, Inc.

A good reporter's job is to ask questions, even if they are not comfortable. Sir Robin Day, the first newscaster of Britain's Independent Television News, recalled the time when he almost caused an international incident.

I was accused of gross discourtesy in an interview with the Japanese foreign minister, Mr. Fujiyama, who visited London in 1957. The extraordinary scene at the Dorchester Hotel, in full view of newspaper reporters and photographers, was none of my making. I had with me a Japanese ball-bearing packet and the original British article of which it was a copy. Pirating of British designs had been in the news as a matter of extreme annoyance to British businessmen. Newspaper reporters raised it at the press conference which Mr. Fujiyama gave immediately before my filmed interview. It seemed only sensible for a TV reporter to produce visual evidence of these practices. I had no thought of creating a rumpus. When I produced the two ball-bearing packets and asked what the Japanese government was going to do about it, the Japanese interpreter was furious.

"This is out of order," he snapped. "If you had this in mind you should have given the minister advance notice."

This was a silly piece of official small-mindedness. The minister had readily answered questions on this subject at his press conference. He would have been quite capable of doing so in the TV interview if the interpreter hadn't butted in. I replied: "The minister did not ask for advance notice of questions." With the ITN camera still turning, poor Mr. Fujiyama sat puzzled and silent between us. The Japanese interpreter exploded: "This is treachery." I could scarcely believe my ears. Keeping as much control of myself as possible, I held up the ball-bearing packets for the interpreter and the camera to see and retorted: "But British manufacturers regard *this* as treachery." (I realized how mild-mannered this retort had been when I met Mr. Aneurin Bevan a few days later. Nye had seen the interview: "You should have said, "Pearl Harbor to you, boy, Pearl Harbor to you.") After two or three minutes another Japanese official realized the whole business was being recorded on film and marched in to stop everything.

The incident had repercussions. It upset the British Foreign Office. A former British ambassador to Tokyo wrote to the Press accusing me of gross discourtesy to a visiting statesman. In Japan a newspaper columnist wrote that Mr. Fujiyama had been "treated spitefully," but he added:

The British have something to complain about. It is shameful to report that Japanese manufacturers have been pirating British designs. Our manufacturers should use their original talents . . . otherwise our ministers will be subjected to stronger criticism when they visit overseas capitals.

There were other consequences of this "discourteous" interview. A week later Mr. John King, managing director of Pollard Bearings Ltd., received a letter from Singapore informing him that Bollard Bearings (the Japanese version of Pollard) were to be withdrawn in that area.

Jesting Pilate

The issue of fairness in documentaries will never go away; some subjects are too controversial to satisfy all partisans. Geoffrey Cox, a New Zealander, produced one of the early television documentary series for Britain's Independent Television, *Roving Report*. He remembered working with an editor on an eminently fair, historical program about the Holy Land.

"Any problems?" asked Cox.

"Yes," his weary program editor replied: "A call from a chap called Pontius Pilate who says his case had not been properly put."

Peter's Pence

During the first papal visit to Jordan and the Holy Land, Paul VI was followed everywhere by mobs of newsmen and cameramen. Adrian Console, one of the top British cameramen, managed to rent a helicopter to cover a large open-air event; he was somewhat astonished to find that his pilot was none other than His Majesty, King Hussein. Apparently, the Hashemite monarch enjoyed both the flying and the extra money. After a day's hard work, he asked Console:

"When will you need me again?"

The pope's visit to the Holy Land fired the imagination of Catholics all over the world. President Kennedy had authorized the minting of a commemorative gold coin, one of which was brought to Jordan by Adrian Console's assistant cameraman, an American Catholic from the Midwest, in the hope of getting it blessed by the pontiff. When the television crew was presented to the pope, the assistant cameraman carefully drew the twenty-dollar gold piece from his pocket and showed it to Paul VI.

"Oh, is that for me?" His Holiness asked.

The cameraman nodded feebly.

"Thank you very much," said the delighted pontiff in perfect English, as he pocketed the donation and walked away with it.

Small Talk

The Today Show scored a first in 1985 by getting Vatican permission to televise a papal mass from the Pauline Chapel during Easter week. Chatting informally with the visiting Americans, Pope John Paul II asked Willard Scott if he was from Philadelphia. The irrepressible weatherman replied:

"No, I'm a Baptist!"

Reach Out and Tape Someone

Larry Glick, hosting a popular midnight show on WBZ in Boston, became legendary for his impromptu phone calls. One of his favorite tricks was the radio version of "candid camera": he engaged people in long conversations without telling them that they were on the air. Finally, the Federal Communications Commission put an end to the practice.

However, Glick met his match when he tried to get the Pope on the telephone. He finally got through to the principal papal secretary, who immediately asked whether the conversation was being recorded. Glick did not want to lie to a representative of God's vicar on earth. "In that case, this conversation is terminated," said the Vatican official firmly and hung up.

On the Spot

Early in her career, television commentator Linda Ellerbee covered the usual daily urban disasters for the 11 o'clock news for WCBS in New York City. She once rushed with her crew to a fireboat reported to be sinking in the East River. The firemen had managed to tie up to another ship, and had evacuated the sinking boat. That was the one Ellerbee and crew boarded. The camera was quickly set up and she waited to go live on the air. Then she asked one of the firemen why their boat did not seem to be sinking.

"Ours isn't," he replied, "but yours is."

Ellerbee tells the story of a news conference following the crash of an airplane in Alaska. A reporter kept pressing the medical examiner to state what precisely had caused the death of 111 people. "Son," the weary coroner told the pestilent newsman, "let me put it this way. The plane stopped and the people didn't."

Unserendipity

Soon after the supersonic Concorde began flying between Europe and New York, Bob Teague was dispatched by WNBC to report on the noise complaints from the residents of Howard Beach in Queens. He was patiently interviewing a mother who described the devastating effect of the jets:

"My nerves are frazzled; my kids can't sleep because of the noise; it's even interfering with my sex life . . ."

Just as she was talking, Teague could hear the unmistakable thunder of an approaching Concorde. What serendipity, he thought, to have the mother's voice drowned out by the airplane. He had almost bagged this perfect ending for the story, when the disgusted voice of the audio man came through his earphone:

"Cut! Sorry, Bob, but that goddamn jet is making too much noise."

It Seemed Like a Good Idea at the Time

Paul Harvey, on his long-running radio column *For What It's Worth*, reported on a local television show in Altoona, Pennsylvania. The anchor of the local news, Brandon Brooks, was doing a report on home protection. He had the bright idea of using his own house as a model, demonstrating exactly where the burglar alarms were located and how the double locks worked.

A couple of thieves were fascinated by the program, and a few evenings later, when they knew that Brooks would be in the studio anchoring his program, the burglars paid a visit to his house and demonstrated both their knowledge and gratitude by cleaning out his house.

How It's Done

During the pre-Christmas shopping season several years back, Heather Bernard was preparing to report on WNBC-TV from Alexander's department store in New York about the various tricks a pickpocket might use to distract a customer. She was about to introduce the store detective to the viewers, when she noticed that somebody had placed her expensive winter coat across the cosmetics counter where it could get smudged by lipstick and makeup. During the few seconds that it took her to grab her coat, Bernard found that her main prop—a wallet stuffed with sixty-eight dollars that she had borrowed from her crew—had been lifted by an enterprising thief who exploited the temporary distraction.

Henry's Choice

Sally Quinn was a star reporter on the Style section of the *Washington Post* when CBS hired her away from an angry Ben Bradlee, her future husband. Quinn was notorious for using her feminine wiles to get subjects to talk. "Maxine Cheshire makes you want to commit murder," Henry Kissinger once told *Post* publisher

Katharine Graham, referring to the paper's gossip columnist, then added: "Sally Quinn makes you want to commit suicide."

Familiar Bedfellows

The late Frank Reynolds, news correspondent and later anchorman for ABC News, was covering the re-election campaign of the governor of a midwestern state. Interviewing the candidate, Reynolds asked what he liked about campaigning. Instead of the usual round of kissing babies and shaking hands, the politician mentioned that he particularly enjoyed having his wife at his side while stumping the state. Without much thinking, Reynolds mused:

"Well, politics, as they say, really does make strange bedfellows."

The governor turned green and, as Reynolds recalled, he did not get re-elected.

Figure It Out

Herman Talmage was governor of Georgia visiting Florida, where a television interviewer asked him:

"Governor, can you tell us how you feel about so many of Georgia's native sons leaving your state for the better life in Florida?"

"I think," Talmage smiled, "that it raises the level of intelligence in both states."

Celebrity Hunt

Bonnie Churchill, whose celebrity interviews are syndicated over twelve hundred radio stations in America, will go anywhere in search of material. She interviewed super-jockey Willie Shoemaker at thirty-five thousand feet up in the air on the way to the Kentucky Derby; she snared Dennis Connors, winner of the America's Cup, in a limousine; and she taped actress Olivia Hussey on a train. "Suddenly we entered a tunnel," the winsome Churchill reminisced, "and I kept asking the questions directed at her sitting in front of me. When we emerged from the pitch blackness, Olivia somehow managed to be sitting behind me, but still attached to her microphone."

Churchill flew once to Mexico to interview Cesar Romero. They were sitting at breakfast at the Las Hadas resort, looking at the menu, when Bonnie confessed that she didn't know how to order.

"Don't look at me," said the epitome of the stereotypical Latin lover, "I don't know this language. I grew up in Brooklyn."

Over more than three decades, Churchill has ventured into some strange nooks in search of quiet and broadcast quality. She recalls making one interview in a broom closet; at another time she was on a golf course, dodging the balls and shooing off gophers. Once she and Stuart Allingham, her indispensable and efficient sound engineer, were taping a program about celebrity moms at Jimmy's. "There we were standing around with Bruce Willis' mother, and Frances Bergen, Candice's mom, in the men's room, the only place where we could find quiet," Stuart remembered. "Fortunately, I borrowed a screen, so the ladies would not be viewing the pissoirs, if you'll pardon the expression."

Ars Longa

Tamás Sipos, a veteran reporter on Hungarian television, went to interview a famous graphic artist, Lipót Herman, on his 75th birthday. He had set up to show the venerable artist in his studio, just as he would begin painting each morning at his easel. When the camera started rolling, the reporter congratulated Herman on being seventy-five and having "already created some wonderful works." Then he paused, waiting for a response. The mild-mannered artist looked a bit startled and then smiled:

"Yes, I thought it was time to get down to work sooner or later."

Way Over the Rainbow

Judy Garland was on *The Tonight Show* discussing with host Jack Paar an actress reputed to have had numerous affairs.

"Isn't she a nymphomaniac?" Paar asked, to which Garland gave what has become a classic reply:

"Only if you can calm her down."

Is That a Yes or a No?

Before an interview with Mae West on BBC radio, her manager asked Gilbert Harding if he could try to sound a bit more sexy. Harding replied:

"If, sir, I was endowed with the power of conveying unlimited sexual attraction through the potency of my voice, I would not be reduced to accepting a miserable pittance from the BBC for interviewing a faded female in a damp basement."

Bosom Friends

Mae West was having some glamour photographs taken while on contract at Paramount Studios. To keep the stars amused during the long sessions, the management had installed a radio which played mood music. It was a hot day, and Miss West felt confined in one of those tight dresses into which she enjoyed pouring her ample body.

"Do you have retouchers?" she asked photographer Whitey Schafer, mischievously liberating one of her milky white breasts, and then pulling out the other. "Can you take care of these?"

It was just at that moment that the host on the radio announced the next pair of singers:

"And now, introducing Mike and Ike!"

Mae West whooped with delight, having just found the perfect names for her two bosom friends.

And Did She?

Baseball star Dizzy Dean was appearing with sex star Betty Grable on a benefit program broadcast from a Veterans' Hospital.

"Say something cute to Miss Grable," the announcer was urging the baseball pitcher.

"Miss Grable," said Dean shyly, "I'll show you my curves, if you'll show me yours."

Nude Radio

An interviewer asked Marilyn Monroe if she had indeed posed for photographs with nothing on.

"Oh, goodness no," the sex goddess replied wide-eyed, "I had the radio on."

Ask Zsa Zsa

Zsa Zsa Gabor moderated a television panel dealing with marital problems. The first guest was on the horns of the following dilemma:

"I'm breaking my engagement to a very rich man who has given me a beautiful house, a sable coat, diamonds, a new kitchen stove, and a Rolls Royce. Miss Gabor, what would you advise me to do?"

Zsa Zsa delved for a moment into her own rich store of life experience and then counseled: "Give back the stove."

CHAPTER 8

Personalities

Knight of the Air

Floyd Gibbons was one of the first newscasters on radio, but he belonged to the world of fiction. In his 1930 portrait of the "Knight of the Air" at his peak, writer Douglas Gilbert gives a sense of Gibbons's ability to dramatize everything, including himself.

He's radio's knight errant; the listener's passport to uncharted harbors; their open sesame to Cathay; their vista of a world whose only boundaries are the poles. The magic of his tales make fireplace dreams come true.

Through him they hear the basso grunts of the trolls in the halls of the mountain kings; peer with him at the sheen of the sunshine silvering ice cap floes; see the finger of a hula crooner twanging a uke as she beckons from Waikiki; hear the adenoidal "Aaaaaaaaaa" of a muezzin calling upon Allah from a desert tower.

"Wait a minute!" said Gibbons, lunging his bulk across the table to peer at these notes, "Knight errant? You mean 'radio's night error.'"

"That's a hot one," said Ed Thorgersen, his announcer, "why don't you use it some night?"

"I did," said Gibbons, "but its illustration was too effective. I bounced into the stage traps of the Great Northern Railway parked in the studio here one night, set train whistles blowing, engine bells ringing and all but ruined four or five programs. It was the first time I ever ran into a train on Fifth avenue—and the last, a couple of production men told me."

Succession

A romantic adventurer in the Dumas *père* tradition, Gibbons could be recognized instantly by the white patch he wore over the eye he lost at Belleau Wood in World War I, his trenchcoat, and a fedora on top of his six-foot frame. His program, *Headline Hunters*, was supposed to be about the news, but dwelled mainly on Floyd Gibbons's adventures with Pancho Villa, General Pershing, and others.

Headline Hunters was sponsored by the publisher of Literary Digest, one of the most fervent backers of Prohibition. Gibbons, whose own convictions lay in another direction, went drinking one night, and on his way back with a number of his boon companions decided to serenade his sponsor. The literary gentleman, awakened to the Falstaffian scene, was not amused and told Gibbons, whom he was paying handsomely, to go home.

Soon afterward, Gibbons lost his job. He was to be replaced by another global traveler, Lowell Thomas, who started his broadcast career with *Headline Hunters*. In his farewell broadcast, Gibbons tried to emphasize the great difference between his own swashbuckling style and the reserved Thomas, obviously hoping that his audiences would miss him all the more after comparing him with an egghead. "I do hope you'll tune him in," he said, introducing his successor, "because . . . Lowell Thomas is one of the world's most profound students of archaeology, philosophy, zoology, and also one of our generation's greatest experts on the dead languages."

Lowell Thomas later recalled in his memoirs that what made him unhappy with this introduction were the omissions: "Not a word did he say about my many years as a news reporter and newspaper editor in Cripple Creek, Denver and Chicago. No mention of my having been a war correspondent in World War I. No reference to my having had the exclusive story of the fall of Jerusalem or how I had scooped every journalist in the world with my account of the Arabian campaign and my exclusive story of Lawrence of Arabia, and also of the German revolution. He was telling his millions of listeners how he, a newspaperman, was now turning over the number one news program to a *professor!*"

Cassandra

H. V. Kaltenborn, the dean of radio commentators in the thirties and forties, was famous for his rapid-fire delivery; he was clocked on the air at between 150 and 175 words a minute, rising occasionally, when he became excited, to 200 words. He spoke fluently without a script, just using a few notes. His ability to provide impromptu and cogent analysis on breaking events amazed his contemporaries. During the Munich crisis in 1938, he made 102 broadcasts in twenty days, translating and analyzing Hitler's speeches along with the latest bulletins that came across the wires. During this period, the sixty-year-old veteran slept in studio 9 at CBS on an army cot, ready to go on the air at a moment's notice. At one point Kaltenborn was so exhausted that when the Archbishop of Canterbury prayed for peace, the commentator began to analyze the homily as if it were a political speech. He became a national figure and was offered to play himself in Frank Capra's movie *Mr. Smith Goes to Washington*.

Despite his celebrity, Kaltenborn's cogent predictions went unheeded in Washington. He warned several times that the Japanese would make a surprise attack on the United States. On December 6, 1941, the eve of Pearl Harbor, Kaltenborn gave his Cassandra-like warning for the last time:

"If we were a totalitarian government we would long ago have issued orders for a sudden surprise attack on Japan. . . . It was a surprise attack on the Russian fleet before war was declared that gave Japan a good start in 1905. But we are a democratic power which waits to be attacked. Which means that we must leave to Japan the great advantage of choosing her own time."

The Lord Is My Shepherd

These days it is not uncommon for defeated politicians to become broadcasters, and occasionally for media personalities to run for office. In 1992, columnist Patrick Buchanan yielded his seat on the CNN program *Crossfire* to John Sununu, who had been pushed out from his job as the White House chief of staff. In 1988, Pat Robertson, the televangelist, ran for president of the United States. And one of the powerful political voices in the troubled 1930s belonged to Father Charles Coughlin, who became known as the Radio Priest.

The Canadian-born Catholic priest had backed Franklin D. Roosevelt in 1932 but gradually turned against the New Deal. Under the influence of the bigoted Dick Richards, who owned WJR in Detroit (and the Detroit Lions), Father Coughlin became increasingly anti-Communist and anti-Semitic. Although he founded a political organization, the National Union for Social Justice, Coughlin did not run for office; rather, he looked for candidates who might carry out his agenda. After Huey Long disappointed him by getting assassinated, Father Coughlin picked William Lemke, a practically unknown congressman from North Dakota to lead the Union Party into the 1936 presidential election against "Franklin Double-Crossing Roosevelt." The Radio Priest was so confident of success that he told his listeners he would disband the Union Party and quit his radio program if William Lemke failed to garner at least nine million votes. He went after FDR with such venom that the Vatican dispatched Cardinal Pacelli, later Pope Pius XII, to rein in the runaway priest.

Following Roosevelt's overwhelming victory and Lemke's failure to gain even one-tenth of the nine million votes, Father Coughlin remained silent for about two months. Then in January 1937, the clerical demagogue took to the airwaves again—reluctantly, he explained—saying that his bishop's dying wish had been for him to resume his ministry on the air.

Whatever It Is, We're Against It

One of the controversial political commentators during World War II was Phillip Keyne-Gordon, broadcasting from Cleveland's WJW over the Mutual Network. He was against drafting fathers, food subsidies, Roosevelt's New Deal, and also opposed Roosevelt's challenger in 1940, Wendell Wilkie. The broadcasts caused even greater controversy when it was revealed that Phillip Keyne-Gordon did not exist; five radio newsmen used the ficticious commentator to vent their views without endangering their jobs.

Britain's Rudest Man

Gilbert Harding, the host of several early quiz shows on the BBC, was a school-master by profession. Partly due to this pedagogic manner and partly to a total lack of unctuousness that seems to infect quizmasters, Harding acquired a reputation for being "Britain's Rudest Man." Whether this was justified or no, Harding claimed that it "provoked extraordinary outbursts of rudeness" against him. Once he was having supper at a private club, before staging a Ladies' Night Brains Trust. One of the women who invited him came up to him and pushed a crumpled piece of paper in front of his nose, asking for an autograph. Harding said he would be happy to oblige after he finished eating.

"Oh, don't think I'm one of your fans," said the woman. "I only want you to sign this for a friend."

"Then why bother," asked Gilbert, "if you dislike me so much?"

"Oh, you nasty rude man!" said the clubwoman.

"Madam," replied the broadcaster, "if I wanted any lessons in rudeness I would go to you for them. Now do you mind if I finish my supper?"

Harding was a lifelong bachelor; he wrote in his memoirs that he might have married Hermione Gingold had the actress been available or willing. The only other woman who had attracted him was a stranger who came up to Harding as he was sitting on a bench with his mother at an English seaside resort. She stopped in front of the broadcaster, and glared at him.

"Ugly man, ugly woman," is all she said.

"Really, madame," Harding inquired, "the ugly man being myself, I presume?"

"Certainly," said the woman.

"And the ugly woman being you?"

"Of course," she said and went on her way. Harding reminisced that "perhaps she had hoped we could have made a true marriage there on the spot."

No Dummy

One of the top radio stars in the late 1930s was made of wood, though most people who heard Charlie McCarthy thought him human. In his book of recollections, None of Your Business, *advertising man and producer Carroll Carroll drew this portrait of Charlie and his less-famous owner.*

One of the most popular rehearsals to attend, often more fun than the show itself, was Edgar Bergen and his dummies, Nelson Eddy (not one of them), Dorothy Lamour, Don Ameche, and the magnificent W. C. Fields putting together a *Chase & Sanborn Coffee Hour*, which also starred personalities from the galactic heights of Hollywood nobility. Glamour girls of the era like Rosalind Russell, Barbara Stanwyck, Bette Davis, Hedy Lamarr, and Lana Turner were always happy to be flirted with by Charlie McCarthy.

Too much has been written about—and I don't think Edgar Bergen likes to be associated with—"the Great Gobbo" syndrome that is reputed to exist between a ventriloquist and his mannequin. But none of the many who attended *Chase & Sanborn* rehearsals could fail to notice that while Edgar made frequent fluffs, Charlie never blew a line. It was Charlie who was always needling Ed about his mistakes. "All right, Bergen! Let's go back and try it again."

When they ran through their routines (See that? *Their* routines! It was Bergen's routine!), Edgar stood in front of the mike while he held Charlie, or Mortimer Snerd, on a specially designed stool placed almost in *front* of the microphone in such a way that the dummy could be plainly seen by the entire audience. So real did Bergen's dolls become that even the audio engineer who rode gain on the show every Sunday would occasionally wigout as he watched Bergen work and holler, "Will someone go out there and tell that dummy McCarthy to stay on mike!"

One Sunday I took my daughter, Leda, to a rehearsal. After the work was over we went onstage. While I was talking to Bergen, Charlie, who was sitting on his stool near the mike, called, "Hey, you. You with the pigtails."

Leda turned toward Charlie and took two tentative steps in his direction.

"You're cute," Charlie said. "Got a piece of gum?"

"No," said Leda grandly. "I don't chew gum."

"Get her," he said in disbelief.

These lines from Charlie came, of course, during *my* half of the conversation with Edgar.

Leda moved a little closer to Charlie. Suddenly he said, "Give us a little kiss."

She stepped back astonished, pulled herself to her full three feet, and said with the air of a mother superior, "I don't kiss wooden little boys."

Charlie just giggled his rakish little giggle. He was never at a loss for words even when Bergen seemed to be out of sight and paying no attention to what was going on where he was. When some curious person would lift the lid to look into the specially built case in which he was carried, if Charlie happened to be inside, the peeker would be greeted with, "Who do you think you're looking at?" Or, "How'd you like someone staring at you when you're resting?"

If someone made a move to touch him, he got the line so often used on W. C. Fields: "Take your hands off me or I'll slug you! So help me, I'll mow you down!"

When it was time to go home and Edgar put Charlie in the case and started to close the lid, if there were any people around Charlie would holler, "Save me! Lemme outta here! This guy's trying to smother me! He's trying to kill me!"

W. C. Fields's phobia of children extended to those of the wooden kind. He genuinely disliked Charlie McCarthy, and once he snarled at the dummy: "I'll slash you into Venetian blinds."

Tribute

The poet Edna St. Vincent Millay was an ardent fan of Charlie McCarthy, and she sent the following parody of nonsense verse to his creator, Edgar Bergen:

> Last night I heard upon the air
> A little man who wasn't there.
> He wasn't there again today;
> I hope he'll never go away.

Night Out

Appearing on Mary Margaret McBride's radio program gossip columnist Hedda Hopper was telling stories of her elderly and somewhat deaf mother's visit to Hollywood:

"One night we went to Ciro's with Edgar Bergen and Ken Murray. Mother ordered milk so I did, too, and the waiters almost dropped dead. Of course, she hadn't caught our hosts' names, but next morning she said, 'Darling, I suppose they were celebrated people we were with last night,' and when I told her she looked blank, so I said, 'You know Charlie McCarthy.' She said, 'Was it now? Why, he doesn't look a bit like he sounds!' You see, she'd heard Charlie McCarthy on the radio, but Ed Bergen never entered her consciousness. She said, 'Ken Murray looked so sad for a funnyman.' I told her, 'All funnymen look sad.'"

The Wrong Eleanor

Mary Margaret McBride, the original chat-show host on radio, was on friendly terms with many celebrities. Once the Girl Scouts of America asked her to invite Eleanor Roosevelt on her program and make an appeal for funds for their organization. Mrs. Roosevelt readily agreed, and McBride left it up to her staff to arrange a broadcast date. The day and hour of the live program came and both broadcaster and the girl scouts were disappointed that the distinguished guest had failed to show up. McBride started her show, explaining just how busy the first lady must be with her projects, daily column, and constant travels. Then her eyes fell upon a familiar, older face in the front row of her studio audience. It was Mrs. Roosevelt, sitting there patiently with a smile. Except she was the first Mrs. Roosevelt, Teddy's widow, whose name also happened to be Eleanor.

Tonight We Improvise

There was so much competition on the airwaves that performers tried hard to come up with a personal signature or trademark. One of the most famous was Bing Crosby's weaving the improvised notes "bub-bub-boo-boo" into his songs. As in most things in Crosby's career, he did not have to try that hard.

It happened during one of his early CBS broadcasts that the sheet of music fell to the floor, and Crosby simply replaced the missing words with those nonsense sounds. The response in the studio was so positive that the crooner repeated it deliberately the next few recording sessions, and then so often that it became his special signature.

It became widely mimicked throughout the land. When Eddie Cantor was doing a charity broadcast for the "March of Dimes," he went down to record a

message from President Roosevelt in Warm Springs, Georgia. Testing the microphone, FDR sang in his own way, "When the blue of the night . . . bub-bub-
bub-boo!"

Songless Detective

The title role of the TV show *Colombo*, the bumbling detective that Peter Falk
has made so uniquely his own, was originally offered to Bing Crosby. The singer
turned it down, allegedly because it would have interfered with his golf.

Laid Back

Early in his career Arthur Godfrey did a morning show in New York on WABC.
Godfrey did not care much for getting up for the six A.M. start of the program, so
it became habitual for whichever staff announcer was on duty to go on the air
with the following sentence: "This is *The Arthur Godfrey Show*. Arthur's not
here." And then he would put on a record until the host showed up.

Diogenes Would Have Liked Him

Arthur Godfrey had no outstanding talents—as a comedian or ukulele player—
yet in the mid-fifties he dominated the airwaves to such an extent that his
combined radio and television audience exceeded eighty million Americans, and
he was responsible for about 12 percent of the revenues of the entire CBS
network. Media critics, like Ben Gross, tried in vain to explain this phenomenon.
Gross polled his readers and concluded: "All I could contribute to the discussion
was a safe generality: he owes his success to 'some element X' in his personality,
which must forever remain a mystery."

   Perhaps a more important element was Arthur Godfrey's straightforward and
natural honesty. He behaved on the air more or less as he was. He kidded
sponsors and politicians, but got on well with the press, knowing what they liked.
He was giving a tour of his new offices to reporters, when he came to its center
piece: a full bar.

   "I built the place for you fellows," Godfrey laughed.

Andy Rooney, now the resident curmudgeon on *60 Minutes*, worked for Arthur
Godfrey as a writer. He selected some of the best stories that fans sent in to the
program for Godfrey to read on the air. Out of this came the idea for a book. But
after Simon and Schuster paid a large advance and published *Stories I Like
to Tell*, they were disappointed when Arthur Godfrey refused to promote the
book, even on his own shows. He told them that he just didn't feel right about

plugging a book of stories that he did not write, and some of which he had not even read.

Catchword

"Good night, Mrs. Calabash, wherever you are," became part of Jimmy Durante's signature. Originally he ended the *Jimmy Durante–Garry Moore Show* in the mid-forties by bidding good night at the end of each program to Garry Moore, then came the mysterious Mrs. Calabash, whose identity was never explained, but she became part of the show.

Jimmy Durante never let on about the identity of Mrs. Calabash, which excited a great deal of speculation about her. Many people claimed to be in the know, including Georgie Jessel, who said that he used to go out with her. According to the producer of the show, Phil Cohan, the name came from a type of pipe smoked by Sherlock Holmes, made from a calabash gourd. Another theory had it that Mrs. Calabash was a racehorse on which Jimmy Durante had lost a great deal of money. Some said that it was what the Schnozzola called his first wife, others that it was a nickname for his partner, Lou Clayton. The second Mrs. Durante, Margie Little, told George Burns that "Mrs. Calabash was the name Jimmy used to mean all the lonely people who were listening to him."

One radio columnist in the early sixties adapted the catchword to Jacqueline Kennedy, who would slip away as often as she could from the formal surroundings of the White House to go riding in Virginia, to see a ballet in New York City, or attend a social event in Palm Beach. Noting her frequent absences, this Washington broadcaster would sign off:

"Good night, Mrs. Kennedy, wherever you are."

Mr. Television

Milton Berle got his title of "Mr. Television" in the late forties, when his Tuesday-night variety show, *Texaco Star Theatre*, on NBC, was credited with persuading millions of American families to buy their own sets. Perhaps a touch of envy made fellow comic Joe E. Lewis comment:

"Berle is responsible for more television sets being sold than anyone else. I sold mine, my father sold his . . ."

Uncle Miltie

Milton Berle himself was responsible for his other epithet, "Uncle Miltie." Viewers frequently asked him to say something to children in the household that would make them better behaved, go to bed earlier, or eat their vegetables. Berle resisted until the opening show of the second season in 1949, which unexpectedly ran shorter than rehearsed—a frequent problem with live shows—leaving the host with seven minutes to fill. After a few ad-libs, which seemed to him like a quarter-hour, floor director Ralph Nelson was still holding up five fingers to indicate that Berle would have to fill another five minutes—an eternity on stage. Finally, in desperation, the comic remembered the parents' requests, so he looked straight into the camera and said:

"Since this is the beginning of a new season, I want to say something to any of you kiddies who should be in bed, getting a good night's rest before school tomorrow. Listen to your Uncle Miltie, and kiss Mommy and Daddy good night and go straight upstairs like good little boys and girls."

To Berle's surprise, the next day strangers ranging from construction workers to little boys were pointing at him saying, "Look, there's Uncle Miltie." The comic started using the tag regularly because, as he wrote in his autobiography, "you can work for years and never come up with a line, a word, a gesture that catches on and becomes identified with you, and without thinking I had lucked into one."

In those early days, Milton Berle owned the air, and his competition knew it. One of them was Bishop Fulton J. Sheen, on the DuMont network, who started one show with this compliment:

"Good evening, this is Uncle Fultie."

Insurance

Milton Berle came to enjoy his role as "Uncle Miltie" and once boasted on his show to Martha Raye that he had millions of little nieces and nephews. "I hope," the film star told him, "your sister has plenty of Blue Cross insurance!"

Low Key, High Tech

Dave Garroway, the first host of *The Today Show*, began on television in 1949 with a half-hour program on Sunday evenings out of Chicago called *Garroway at Large*. The host wanted to end the first show in a decisive manner by taking an ax to the coaxial cable connecting the camera to the world at large. The engineers "gave kind of an engineering laugh," thinking that Garroway was kidding, but in fact that is exactly what he did, very definitely ending the show.

Unlike the stage, or even radio, television demands the least amount of projection, which stumps many a performer used to live audiences. When in 1952

Dave Garroway began hosting *The Today Show* on NBC, he was so low-key that people were asking:

"But what does he do?"

"Nothing," so the reply went, "but he does it so well."

Mugging

When Dave Garroway started hosting *The Today Show*, there were few indications that the morning program would become an American institution, still running forty years later. To counteract Garroway's low-key, professorial image, a series of "Today Girls" were introduced; they included actress Estelle Parsons and Barbara Walters. The program was still limping along until the producers hired a young chimpanzee named J. Fred Muggs, as co-host. His antics made the ratings soar, and Muggs ruled the set for four seasons. "You could put a baboon on television," as anchorman David Brinkley once observed, "and he'd become a celebrity if you kept him on long enough."

Dave Garroway never felt comfortable with Muggs, who frequently attacked him. There were times when the chimpanzee's bites would draw blood, and Garroway, like a good trouper, kept the show going by turning his other cheek to the camera. Muggs grew meaner with his celebrity status, and the NBC infirmary administered some twenty tetanus shots against chimp bites.

On Garroway's complaints, NBC canceled Muggs's contract, issuing a press release that the primate had left "to extend his personal horizons"—a general euphemism in the broadcast industry for such situations. Then the chimpanzee's owners added insult to injury by suing Garroway for half a million dollars. The case dragged on for several years, and finally the network settled with Muggs for a few thousand dollars. And in January, 1977, when *The Today Show* held its silver anniversary show, J. Fred Muggs was one of the few past participants who did not receive an invitation.

Early Merv

Singer Merv Griffin became one of the most prominent and richest entertainers in American television. The beginning of his career promised little of what was to come:

"I went into television game shows," Griffin once recalled, "just after the game show scandals broke and everybody was going to jail. I became Jack Paar's substitute host, but when he quit Johnny Carson got the job. . . . The first time I substituted for Jack Paar, I did a monologue and walked off the stage. I was about to keep going out the door. The producer stopped me. 'Merv,' he yelled, 'you can't leave. We've got eighty more minutes to fill.' And I said, 'But I bombed out there. I can't do it.' He pushed me back on stage and I've been there, facing those cameras, ever since."

The Proof Is Plain to See

Never having finished high school was often on Jack Paar's mind, and he frequently reminded people about his lack of education. After he had heard the line once too often, Oscar Levant, composer and wit, told him:

"Jack, I'm willing to take you at your word."

Forbidden Fruit

Carmen Miranda's father did not want her to go on stage, so she sneaked off to sing secretly on the radio. Her father was listening one day and said: "That girl is a very good singer," not even recognizing his daughter's voice. She became famous for carrying a bowl of fruit on her head, which she did even on radio, as Peter Duncan recalled her appearance on his BBC program, In Town Tonight, *from the forties.*

When she came into the studio for her broadcast, she was wearing five-inch platform heel shoes, and an enormous display of fruit on her head—apples, oranges, pears, bananas.

"Where did you get the idea for those hats?" we asked.

Carmen grinned. "In Brazil the girls carry fruit in little wooden bowls on their heads. I think I would like a hat like that, so I design them with all the different kinds of fruit I can find."

"They look good enough to eat," I remarked, glancing up at her "salad."

A few weeks later I went to the Savoy to say goodbye to Carmen and her film-director husband, Dave Sebastian. She was wearing the same concoction on her head. This time, grinning all over his face, Dave pointed to it and said: "Have a banana?"

Rather gingerly, I reached up and touched one. It was real.

"Go on," Dave said. "Take it."

I plucked it off, skinned and ate it. It tasted good.

"What about an apple?" I asked next.

Carmen gave me that delicious grin that seems to spread from ear to ear.

"We think you are hungry." She patted my thirteen stones. "You are not fat enough. We put the banana there specially for you. The apple . . ." She shuddered. "That is for me—but not to eat."

Over the Sound Barrier

Lasting only one season in 1963/4, *The Judy Garland Show* is still remembered for the extraordinary range of personalities who flocked to appear with Judy. Many of the great stars from MGM's glorious past—such as Gene Kelly, Mickey Rooney, Jane Powell, Ray Bolger, Lena Horne, and June Allyson—came to chat, kid about, and perform. Garland also kept her eyes out for fresh talent, and she became a fan of a young and rising star by the name of Barbra Streisand. The two singers made a rousing medley out of "Get Happy" and "Happy Days Are Here Again" and were chatting away in a semi-improvised segment called "Tea for Two." Suddenly, out of the studio audience, a familiar voice started belting out "You're Just in Love." It was Ethel Merman, who said she happened to be taping next door an episode of *The Red Skelton Show*.

"Were you belting in there?" Judy Garland asked the reigning queen of Broadway.

"I was belting?" Merman pretended to be indignant. "I heard the noise, and that's why I came in! I just came in to say hello, that's all."

Inevitably, the two legends and one legend-to-be ended up belting out Irving Berlin's "There's No Business Like Show Business," which was Merman's signature song. "The performance stopped the show," wrote Coyne Steven Sanders in his book *Rainbow's End;* "one of the great moments in television and one of the great moments in show-business history."

And at the time *Variety* commented, half in earnest, on the decibel level achieved by the three belting ladies: "The CBS soundmen, incidentally, are threatening to sue Judy, Ethel, and Barbra for bustin' their eardrums."

Tallulah

Although a confident performer on stage, Tallulah Bankhead's larger-than-life personality did not work on the more intimate medium of radio or television. She was apt to be aspish, and Groucho Marx called Tallulah's on-air technique "the timing of the shrew."

Tallulah had altercations with some of the great comics, and usually she came off worse. Once Bankhead invited Bob Hope as a guest on her radio show. When Hope came on too early, she snarled:

"Get off this stage, until I call for you."

"Don't lower your voice to me," Hope snapped back, "I knew you when you were Louis Calhern."

When Bankhead tried to sing at Jimmy Durante, he suggested:

"I think you ought to have your tonsils out."

"I've already had them out," Tallulah said.

"Then put 'em back again," the comedian topped her.

Not Jerry's Kids

Tallulah Bankhead was extremely nervous in preparing to host *The Big Show* on NBC radio in late 1950. Media critic John Crosby, reviewing her debut in generally admiring terms, called her "an unpredictable volcano who has been known to sweep away whole villages when she erupts." He described her singing voice as having "more timbre than Yellowstone National Park," and claimed to have counted either 422 or 425 uses of Tallulah's trademark "dahling" during those ninety minutes.

Although the show became a success on Sunday nights, it was not easy to come up each week with big-name guests, which was the premise behind *The Big Show*. Jerry Lewis was becoming hot just then, and Bankhead found out that the young comic had given her staff—whom she usually called "my boys"—a hard time in negotiating to come on the show. She was further infuriated when Lewis kept the whole cast waiting for the first rehearsal; he swaggered into the studio without a shirt and with his electric shaver, as a visual explanation for being late. La Bankhead sailed up to him script in hand, and whacked the comedian quite hard on each cheek.

"That," she told him, "is for treating my boys so badly."

Ocular Evidence

After one of her last appearances on his show, Merv Griffin asked Tallulah Bankhead why she bothered to appear on television. The actress replied:

"It's one way of proving to people that I'm not dead."

Three Graces

To demonstrate public spirit, commercial television shows sometimes get behind a safe political issue. A favorite, perhaps because it comes around only every four years, is to inspire the electorate to leave the couch and stop watching television long enough to get to a voting booth. Following the great quiz show scandals of 1959, some surviving programs got behind a Get Out the Vote campaign. Chester Feldman, producer of *I've Got a Secret*, invited three heavyweight ladies of the theatre—Tallulah Bankhead, Lauren Bacall, and Helen Hayes—to do a series of public-service spots during the live broadcast of the game show.

Tallulah Bankhead, daughter of a congressman was ardently public-spirited; in fact, she was fond of spirits of any description. Just before the broadcast was to begin from CBS Playhouse 50, the wardrobe lady came rather anxiously to Chester Feldman with the discovery of a quart of Wild Turkey bourbon among Miss Bankhead's effects. Feldman held an urgent discussion with Gil Fates, the writer of the show (and who told this story in his memoirs); they concluded that since Bankhead had managed a forty-year career with her notorious drinking and other outrages, they would be better off doing nothing.

A little later, the wardrobe mistress returned with news that she had been asked by Lauren Bacall if someone could run out and get a bottle of rum for her. Another brief conference was held, and the producer and writer decided to comply with Miss Bacall's wishes.

Miss Helen Hayes was the third of the *grandes dames* to arrive. After Chester Feldman went to greet the First Lady of the American Theatre, he ran into Gil Fates who asked:

"Everything O.K.?"

Feldman announced with a completely straight face:

"At least Miss Hayes had the decency to arrive drunk."

How to Get Your Own Way

Among the first assignments received by Kenneth Whelan, a tap dancer who became one of the pioneer directors in American television, was to make a one-minute promotional spot for a program starring Helen Hayes. The diminutive actress was a giant in the American theatre, an Academy Award winner, and young Whelan decided that this spot would be an opportunity to establish himself in front of his coworkers and superiors, by telling Miss Hayes exactly how she should act. The actress listened and did exactly as she was told for two takes. It was obvious, even to the director, that his approach was not working. He began to sweat and did not know what to do. He called a five-minute break, tried another take, but the result was still terrible. Completely at his wits' end, the young man heard Helen Hayes address him in a very clear, loud voice, to make sure that everybody heard her:

"Ken, you're being very patient with me, and I appreciate it. I think your ideas are great, but I just can't seem to do it in the right way. I'm an old war-horse who is very set in her ways. . . . I have my bag of tricks, and I guess I'm too old to learn any new tricks. . . . Would you indulge an old lady, and let me try it my way?"

Whelan gratefully accepted, and naturally her way turned out to be exactly right. As she was leaving the studio, the First Lady of the American stage turned to the young director, and said again in a very loud voice: "You're a charming and talented director. Thank you for being so kind."

Phoning It In

In 1937 David Sarnoff became obsessed with the idea of bringing Arturo Toscanini back to New York to build an orchestra just for NBC. Unfortunately, only the year before the Italian maestro had abruptly resigned from the helm of the New York Philharmonic and vowed never to come back to the United States. Employing the music critic Samuel Chotzinoff as an intermediary, offering a great deal of money and total control, Sarnoff gradually managed to persuade the seventy-year-old conductor to undertake a series of ten concerts, which stretched to a unique partnership of seventeen years.

To spare the maestro unnecessary travel, NBC hired Artur Rodzinski as an assistant director, and shortwave links were set up between his Italian residence in Milan and Studio 8-H at the Rockefeller Plaza, which allowed Toscanini to monitor rehearsals.

Finally, the maestro arrived for the first Christmas concert, and his boat was greeted by both David Sarnoff and Samuel Chotzinoff.

"NBC orchestra very good," were Toscanini's first words; "first clarinetist not so good."

That Long Distance Feeling

Rehearsing a Mozart symphony with the NBC Symphony Orchestra, Arturo Toscanini asked the first violins to produce a "faraway effect." They tried, but the conductor said:

"Too far away." They tried it again, and Toscanini rapped his baton: "Not far away enough."

The violinists were frustrated, and one of them asked:

"Maestro, will you please tell us how far away you want it?"

"I want it just far enough," said Toscanini, "you know, like Brooklyn."

My Musical Career

Benjamin Kubelsky's father bought him his first violin when the boy was six years old, paying a whole year's savings for it. The family kept lavishing its small discretionary income on private lessons for Benjamin. He was kicked out from his high-school orchestra in Waukegan, Illinois, while rehearsing Schubert's *Unfinished Symphony*; young Benjamin remarked that the orchestra would certainly *finish* it. The following year Benny was expelled from the school itself.

Jack Benny felt guilty for the rest of his life for having become a comedian rather than a virtuoso. And beyond admiration he always envied the great violinists, even when he had one of the top-rated shows in radio and then television.

Benny was awestruck by great violinists, and considered the highest honor to be appearing on the same program with Isaac Stern, or just to be in the presence of a Szigeti or Heifetz. The latter did compliment him once on his rich tone, and thought it was a pity that he had not pursued music more vigorously as a child.

"A guy like Heifetz," Benny would complain, "doesn't have to drive himself crazy to have them compose eight numbers for his next concert. He gives the audience the good old Beethoven and Paganini numbers. Also Heifetz doesn't have to worry about his Nielsen rating and about CBS vice presidents and agency men or sponsors. I should have listened to my mother and practiced harder when I was a kid."

Although he mainly used his violin as a weapon to deflate his staged self-esteem, Jack Benny was quite a skillful violinist, who often gave benefit performances for charity. Once, to help the Greek War Relief, he gave a virtuoso performance of a concerto arrangement of "Love on Bloom." After he performed all the tricky staccatos and arpeggios, a stunned audience gave Benny a huge burst of applause, and someone he knew came up to him:

"Jack, I never realized you played the fiddle so beautifully."

"Listen," Benny kept his usual mournful expression, "when I was younger they used to call me another Heifetz."

A pause and a sigh: "Not Jascha—*another* Heifetz."

Feuds

Fred Allen was asked once by *Screen Guide* to review one of Jack Benny's films. "Mr. Benny first appears in a swing band," wrote Allen, "although he couldn't get a job wetting the finger of the man who turns the music for the washboard player with the Hoosier Hot Shots."

At a Friars' dinner honoring Jack Benny, Fred Allen made the famous comment on his feuding partner's talent with the violin: "Jack is the only fiddler who makes you feel the strings would sound better back in the cat."

Ed Sullivan feuded with just about everybody, most notably with Walter Winchell, which ended in enduring hostility. He also accepted a challenge in 1961 to debate Jack Paar on *The Tonight Show*, which the media hyped into an event almost on par with the Kennedy-Nixon debate the year before. It fizzled when Sullivan bowed out a few hours before the show, claiming that Paar wanted to change the rules. That left Paar to fill the void with a fifteen-minute monologue in which he belabored his absent opponent for lying:

"Ed Sullivan has proved to be as honest as he is talented. . . . He is afraid to appear on the show—not that he would be murdered, but that he would commit suicide in front of an audience. . . . Everybody looks more interesting here, and if you had been here, Ed, you might have looked more interesting for the first time."

Mr. Rigor Mortis

More barbs were hurled at Ed Sullivan and his show, *The Toast of the Town* (frequently parodied as "The Toast of the Tomb"), than perhaps at any other television entertainer. Sullivan's enduring success, despite his stiff posture, unsmiling expression, awkward mannerisms, and frequent fluffs, greatly puzzled some of his fans, but especially insiders. Fred Allen, not given to professional animosity or jealousy, seemed genuinely offended by Sullivan's lack of talent. "He's a pointer," he once said, "and you can teach a dog to point. Just rub meat on the target."

Sullivan, who had an audience of 35 million viewers, retorted: "Maybe Fred should rub some meat on his sponsor."

Fred Allen hit back at the stony-faced master of ceremonies: "Ed Sullivan will be a success as long as other people have talent."

"Ed does nothing," Alan King marveled once, "but he does it better than anyone else on television."

And Henny Youngman remarked: "In Africa the cannibals adored him: they thought he was some kind of frozen food."

Apart from his awkward stiffness during his long years on television, Sullivan was mainly known for his stony-faced expression that rarely varied. Alan King says the best line he ever heard about the grim-visaged emcee came from fellow comic Joe E. Lewis, who once said that "Ed was the only man who lit up a room by leaving it."

One of the few things that used to make poker-faced Ed Sullivan laugh was a recollection of his visit to County Cork, Ireland, where his ancestors hailed from. He was enjoying a hot cuppa at Bantry Bay, proudly telling a waitress:

"I am the first of the clan to get back here."

"Is that so?" the woman responded. "Well, it took ye a damned long time!"

You Can Go Home Again

Like many a person in and out of show business, Johnny Carson earned some money by ushering for a local movie theatre in Norfolk, Nebraska. In 1976

Carson was invited back to give the commencement address at his old high school.

"In case anyone wonders why I came back to Norfolk," the talk-show host began, "I've come to find out what's on the seniors' minds and, more important, to see if they've changed the movie at the Granada Theatre."

Genesis

During a question-and-answer period with viewers, Johnny Carson was asked:
"What made you a star?"
"I started out in a gaseous state," Carson replied, "and then I cooled."

Off the Cuff

Although he uses writers (whom he also blames for jokes that fall flat) Johnny Carson is also well known for his off-the-cuff comments. Once Mr. Universe was lecturing him about the importance of fitness.

"Don't forget, Mr. Carson, your body is the only home you will ever have."

"Yes, my home is pretty messy," Carson ad-libbed, "but I have a woman who comes in once a week."

At another time Carson asked Fernando Lamas what made him go into the movies.

"Because it was a great way to meet broads."

"Nietzsche couldn't have put it more succinctly," said Johnny.

Epitaph

In 1977 Harvard's Hasty Pudding Club nominated Johnny Carson as its Man of the Year. In the press interview before the award ceremonies, someone asked:
"What would you like your epitaph to be?"
Carson paused for a moment, and said:
"I'll be right back."

Typical Fans, Average Guests

A week or so before he succeeded Johnny Carson as the host of *The Tonight Show*, Jay Leno was a guest himself on *The Oprah Winfrey Show*. Afterward he came through the green room, where he ran into four men. They congratulated the comic on his new job, asked for his autograph, shook his hand, and warmly wished him the best of luck. Leno left them, tingling from all the camaraderie, but when he reached the parking lot, he felt sorry that he had not found out more about his newfound fans.

"Who were these guys?" he asked one of the studio attendants.

"Oh, they're Oprah's next guests. Men who murdered their wives."

CHAPTER 9

Die Laughing

Stop, Thief!

Bert Wheeler, an old vaudevillian, had a hard time adapting to radio and then to television. "In the old days," he remarked in the early fifties, "I could buy a complete vaudeville routine for five hundred bucks and use it for five years without changing a line, and with no other comic ever stealing it. Today you pay a thousand bucks for a sketch that lasts one television program—that is if a friend hasn't dropped by for rehearsal and beats you to the airwaves with it."

Or as George Burns has said: "The biggest adjustment we all had to make between vaudeville and radio was that in vaudeville seventeen minutes of good material could last for years, while on radio seventeen minutes of good material would last seventeen minutes."

Walter Winchell used to call Milton Berle "The Thief of Bad Gags," and there were psychoanalytical theories that he acquired the habit because his mother, who had worked as a store detective, once had been scared by a shoplifter.

Berle used to reply to accusations of thievery: "No gag is new until it is forgotten."

Replacing Rudy Vallee on a radio show, Milton Berle greeted him on the way into the recording studio with, "Your time is my time." On another occasion, Rudy Vallee complained: "Fred Allen is always stealing my material: I do a joke on the air, and he does it the week before."

Milton Berle, who claims to have a ten-million file of gags and jokes, enjoys his reputation—even as a thief. He had himself introduced once as "The man who always steals the show—one joke at a time." One of his frequent gags in the old days ran: "I went to see Frank Fay at the Palace. He was so funny I almost dropped my pad and pencil."

One time Bob Hope accused Milton Berle of stealing all his radio jokes. As Berle recalls, his mother, the formidable Sara Berlinger, got very upset and declared: "My son would never stoop so low. My son stoops high!"

Inspired Text

During an award ceremony, comedian Garry Moore was praised for his spontaneity, which he generously attributed to the four writers on his show. Bishop Fulton Sheen was the next to come to the rostrum to acknowledge his award for broadcasting.

"I also want to pay tribute to my four writers," he said, "Matthew, Mark, Luke and John."

Ad-lib

Jack Benny, who appeared as one of the most natural performers, very rarely permitted ad-libs. Everything had to be scripted. Harry Conn, a top-paid writer, said after leaving his employ: "Benny couldn't ad-lib a belch after a Hungarian dinner."

Jack Benny summoned once his whole team of writers to his dressing room, and told the expectant group that he needed the following line inserted at the end of the show: "Professor LeBlanc was played by Mel Blanc."

There was a long pause, and then Milt Josefsberg, the head writer at the time, said:

"You know, Jack, just two of us could have handled this."

The Miser

Jack Benny started on radio in 1932, when his career was laid waste by the Depression and the death of vaudeville. It took him five years to build what someone described as "the most complete and fully realized character ever achieved by a comedian."

Jack Benny's act succeeded largely because it seemed so natural; there seemed to be no act at all. However, many listeners assumed that Benny was as stingy as his famous on-air persona. A lawyer once wrote from Cleveland, berating the entertainer for being always in arrears for Rochester's salary, his valet on the show, and paying him so little in the first place.

Benny shrugged off the letter, but more kept arriving, acccusing him of exploitation and racism, of being too sheltered from the struggles of ordinary mortals. The comic ignored these missives, too, until the lawyer wrote directly to Rochester.

"Please stop writing those foolish letters," Benny finally replied. "They're giving me a pain in the neck. I only hope you're making in one year what Rochester makes in a month."

Eddie Anderson, the black actor who played Rochester, was earning $2000 a week at the time; two servants took care of his needs in his mansion; he owned three cars and several race horses.

"Why doesn't a man like that," Benny commented on the lawyer to journalist Maurice Zolotow, "have the brains to figure out that if I was really such a bastard like I'm supposed to be on my program—would I broadcast it to the whole world?"

Now Ain't That Sumpin'?

Amos 'n' Andy was the first overwhelming success in radio comedy, remaining on the air for more than two decades. But, as Redd Foxx tells it in his book about Black Humor, the show had a curious history. The original characters were created by the well-known and inventive vaudeville team of Aubrey Lyles and Flournoy Miller. Although most of their work was done on stage, Miller and Lyles also produced a radio show out of Chicago, which was sponsored by Log Cabin Syrup. Like many of the traditional minstrels, Miller and Lyles—even though they were black—blacked their faces and spoke in an exaggerated black comic dialect. The show attracted the notice of CBS, which wanted to develop a national broadcast. But when the sponsors and the network realized that they were dealing with black performers, they dropped Miller and Lyles. Broadcasters in those so-called golden years of a segregated America were afraid to offend white audiences by bringing black voices into their homes. So there were anomalies, such as Beulah, the black maid on the long-running radio comedy *Fibber McGee and Molly*, being played by Marlin Hurt, a white man. It all began when CBS decided to create a black minstrel show with white minstrels.

The producers were impressed by Charles Correll and Freeman Godsen, a couple of white Chicago vaudevillians, who based their audition on favorite bits from Miller and Lyles acts, which they happened to love. They began a show called *Sam and Henry* on WGN in January 1926; two years later they moved to WMAQ but found that WGN owned the names of their characters; so *Amos 'n' Andy* was launched.

Meanwhile, Miller and Lyles continued to work in vaudeville, but every attempt to launch a radio show, now based on the proven formula of *Amos 'n' Andy*, failed to leap over the color bar. Many people in show business were aware of the theft. Mark Hellinger, the Broadway columnist, wrote in 1930:

"For more years than I can remember, I can recall Miller and Lyles using lines like 'I is regusted.' 'It all depends on da sitcha-ation you is in,' and similar construction, so similar in fact that the resemblance is uncanny. If Miller and Lyles were substituted for Amos and Andy one night, I would defy anyone to tell the difference. All the credit in the world to Amos and Andy. Ever since last August they have had this nation in hysterics. When I tune in on them and see my neighbors going into fits of laughter, I often wonder what Miller and Lyles are thinking about the success of Amos and Andy."

Aubrey Lyles died a few years later, but Flournoy Miller continued to work on Broadway and then moved to Hollywood where he became the head writer of—what else?—*The Amos 'n' Andy Show*. He continued to work on the television version, which was launched in 1948. By then American society had changed to some extent, and the public would not buy that their favorite black characters were white. Too old to perform his old character again, at least Miller had the satisfaction of finding such black performers as Tim Moore in casting the show.

Pumping Gas

Will Rogers had a problem at first transposing his folksy style to radio. On stage he employed various props, such as his lasso, to break up his monologues. The microphone reminded him of an automobile radiator cap. "I was afraid the thing was going to bite me." He felt better after he started using an alarm clock as a prop. He introduced it, he said, to wake up listeners who had fallen asleep during his monologue.

Rogers wrote all his own material, rather than hire gagmen. When a famous gagman offered to write his show for a thousand dollars a week, Rogers simply reciprocated the offer.

The Gulf Oil Company sponsored *The Gulf Show*, Will Rogers's radio program on Sunday evenings, between 1933 and his death in 1935.

"I don't know whether I am any good until I see whether they have sold any gas or not," Rogers explained to his audience. "Imagine some performer standing up in front of these cemetery associations, see, and he don't know how funny he is until he finds out how many people died and bought lots that week."

There Is Always Hope

As another vaudeville performer used to live audiences, Bob Hope was nervous talking to a microphone. In his early days on *The Rudy Vallee Show*, engineers were mystified by a regular thumping sound that occurred only during Hope's routine. Then they watched his act closely in the studio and found that the comic was kicking the mike with each punchline.

It took Bob Hope a while to master radio, and when he got his first show, sponsored by Pepsodent, the comedian left little to chance. Realizing that the medium had a bottomless appetite for new material, he hired a lot of young eager writers at low wages. Hope drove them hard, asking each one to write a whole show and then selected the best jokes from each. He rehearsed a script that was two or three times longer than the half-hour he needed on the air, and tested it before a live audience. The hallmark Hope monologue has been honed through more than six decades of vaudeville, radio, television and after-dinner speeches. He developed the technique of packing his monologues with 23 compressed jokes or one-liners delivered roughly every twenty seconds, and sometimes twice as many.

Hope cracked so many jokes so fast that some of the weaker ones could whiz by almost unnoticed. But he also bought an insurance policy and rescued a bad joke by sharing the problem with his audience, a technique now made very familiar by Johnny Carson and other television comedians. When a line crashed like a lead balloon, Bob Hope would take a self-deprecating bow and ad-lib: "Move over and let me in on the wake." Sometimes, he might tear the whole page from the script and throw it away.

Bob Hope exhausted his writers, who had to be on call at all hours, on weekends and holidays. Even during the summer breaks from his shows, Hope often hired part-time writers. One of them approached Norman Sullivan, a regular gagwriter on Hope's team:

"What does Bob do about summer?" Sullivan looked at him dolefully and said: "He lets it come."

Blooming Narcissus

Milton Berle dominated every aspect of his television show; by all accounts, except his own, he was an absolute tyrant on the set. George Burns tells a story about the comedy writer Larry Gelbart and his wife visiting a rehearsal. They watched for some time Uncle Miltie telling people what to do and yelling at everybody. It was unpleasant for them since socially they were good friends with the Berles. As they were leaving, Gelbart turned to Hal Kanter, a veteran writer working on the show:

"Let me ask you a question. After watching what went on here all day, there's something I just don't understand. How can a man with your track record put up with all the crap he gives you?"

"I do it," Kanter responded, "because he's paying me five thousand dollars a week. Now let me ask you a question. Why are you his friend for free?"

Renowned for his egoism, Milton Berle had a difficult time with guest stars on his show. He resented even an atom of the audience's attention being wasted on anybody else. He would visibly fidget and twitch while a guest performer was doing his act, and in some extreme circumstances, Berle grabbed the person by the lapel and forcibly turned him away from the camera. Jackie Gleason had prepared himself just for such an eventuality, by putting straight pins under his lapels. "It didn't bother Milton at all," George Burns recalls in his book *All My Best Friends*. "He just screamed and turned Gleason around."

The Perfect Fool Meets Funny Girl

In 1949 Ed Wynn starred in a half-hour comedy show on CBS, one of the first to originate from Hollywood. In those early, experimental days of television, even seasoned comics lacked self-confidence in the new medium. Hal Kanter, who helped write the show, remembers the sense of relief that everybody experienced after the gala premiere of the first episode. The entire production team was waiting backstage for Ed Wynn to emerge from his dressing room and receive their plaudits. He was delayed by a long phone call, after which he came on glowing:

"Well, gentlemen, that's it. We're a smash, that's the definitive word. Do you know who that was on the telephone? That was Fanny Brice herself."

Apparently, the original Funny Girl had watched the show, and in Ed Wynn's opinion she was the greatest American theatrical star. After the second week, she called again to congratulate, which again transported Wynn into seventh heaven. This mutual admiration club went on for several weeks.

After a while the problem arose of finding big-name guests, on whom the show depended. Most of the established Hollywood stars avoided television, because it was giving away entertainment for free, or their studios expressly forbade them to appear in the rival medium. One day, the writers were discussing this problem after a show, while waiting for Ed Wynn, who was taking yet another adulatory call from Fanny Brice. When he came to join his team, one of the writers, Seaman Jacobs, suddenly had a brainwave.

"Hey, what are we all standing around here talking about a star for? There's your star right there!"

"Who?" Wynn asked.

"Why, Fanny Brice!" Jacobs enthused.

"Fanny Brice?" The great clown fixed him with his doleful stare, as he savored the name of the woman whom he had been calling the greatest talent in show business. Then he asked:

"What the hell can she do?"

They Liked Father, They Liked Son

Ed Wynn's television show was canceled, and apart from a few guest appearances, the great comic who had been in show business for half a century, and in over twenty Broadway shows, could not find work. "I've been around longer than RCA or General Motors," he would say to old friends like George Burns. "I'm a brand name, like Heinz 57."

At first the comic refused to change his carefully built-up persona of "The Perfect Fool." But when he came to be known chiefly as Keenan Wynn's father, Ed Wynn was forced—at the age of seventy—to accept even serious roles on television. His most famous one was the part of Jack Palance's manager in the 1956 drama by Rod Serling, *Requiem for a Heavyweight*. During rehearsals, Wynn played for laughs, reaching deep into his bag of vaudeville tricks. When director Martin Manulis asked him to cry, Wynn tried and said he could not. Manulis told him to go home and practice in front of the mirror. The next day Wynn came back and said that every time he looked at himself in the mirror, he got hysterical.

Both the director and Rod Serling wanted the old comedian replaced, but with a great deal of coaching from his son, Ed Wynn played the role to perfection and received an Emmy award as Best Supporting Actor. After the live broadcast, Keenan Wynn declared proudly: "Here I am, Ed Wynn's son all over again."

"There's something wrong with television," Ed Wynn told columnist James Bacon shortly before his death. "What kind of medium is it that has Dinah Shore getting laughs and Ed Wynn making people cry?"

Comic Relief

Harpo Marx, unlike his brother Groucho, had difficulties adapting his special brand of comedy to the electronic media. A mime tends to be a one-joke phenomenon on radio, and Harpo was trapped in his long-time persona, unable to change. When Ed Murrow brought television cameras to the Marx home for an episode of the celebrity program, *Person to Person*, he opened:

"I hope it's not your intention to monopolize the conversation this evening."

Later, in the 1960s, Harpo played a few straight roles on television. "You don't know what a relief it is," the silent comic observed to George Burns, "not to have to get laughs."

Murder, They Wrote

Steve Allen has predicted that Sid Caesar's two consecutive variety series, *Your Show of Shows* and *Caesar's Hour*, will never be equaled as the greatest comic achievements on television. These shows in the fifties are credited (or debited) with killing off Broadway, even though the writers on the show became some of the greatest entertainers on Broadway and in Hollywood: Mel Brooks, Carl Reiner, Larry Gelbart, Danny and Neil Simon, Woody Allen, and others.

Mel Tolkin, who was Caesar's head writer, remembers the murderous competition among the writers to get their skits on the air. At the noisy story conferences, "Doc" Simon, who became the most famous Broadway playwright of his generation, was so intimidated that he would whisper his ideas to his brother or whoever was sitting next to him, because he could not utter them aloud. Woody Allen, enormously fertile with jokes on paper, was equally shy.

Not surprisingly, most of these writers were in therapy at the time, and they also worked out their hostility and anger at each other through the sketches. "I realized later that we got a lot of laughs out of murders," Mel Tolkin recalls. "In one Italian-movie satire, for example, Sid stabs his wife to death, and then, while he's mourning and carrying on, a shoeshine boy comes in to polish his shoes. Almost absent-mindedly, Sid plunges the knife in the kid's back. Then, after he's arrested, Sid walks past the morgue attendants carrying off the bodies, and he also matter-of-factly knifes one of them."

Play It Again, Sid

Among the most successful components of Sid Caesar's shows—with lasting influence on the films that Mel Brooks and Woody Allen would go on to make—were the parodies of current hit movies: "Aggravation Boulevard" or "A Trolleycar Named Desire." Some of these sketches went on for half an hour, using intricate sets and comic business, all live. Producer Sam Spiegel threatened to sue when he heard that *Your Show of Shows* was about to burlesque *On the Waterfront*. But when he saw the result, he was so delighted that he asked Sid Caesar to run it again when the film was being re-released. And he was reported to have sent a script of his upcoming *Bridge on the River Kwai* to Cacsar so that he could parody it all the better.

Smartass

Woody Allen was generally liked by the crazy writers who worked for Sid Caesar. They felt protective toward the "kid" who looked even younger than his years. Danny Simon took him under his wing. But Allen's humor sometimes landed him in trouble, especially with people who lacked a sense of humor. One of them was Max Liebman, the powerful producer of many of the early television comedy shows. When Woody Allen was summoned to his first interview, he could barely see the Viennese-born producer behind his huge desk, covered with three dozen awards, ranging from Emmys to Pcabodys. Allen took in the forest of statuettes and said:

"Gee, Max, I didn't know you played tennis."

It was the end of the interview, though Liebman hired Allen not long afterward to work on a show with Buddy Hackett.

My Son, the Writer

Hal Block, who wrote for many successful comics, including Abbott and Costello, wanted to demonstrate to his father that he had made good, even though he quit law school against the old man's advice. He thought that finally his father would agree with his decision when he saw the extent of his success. After Block senior sat through one of the best radio shows his son had written, heard the wild applause of the studio audience, he said to the expectant Hal:

"Well, now are you ready to go back to law school?"

Kosher Specifications

There is a saying among television writers: "Do you want it good, or d'you want it Thursday?" Remarking on how most comedy writers on radio happened to be Jewish, George Burns tells the story of an advertising executive who sent a memo to a producer, advising him to hire a bunch of young writers for a particular show.

"I'll be glad to hire as many young writers as you like," the producer replied, "but if you want the scripts done on time, I also need two old Jews and a typewriter."

Youth Is Wasted on the Young

Television is a youth-oriented medium, run by relatively young people. Not just actors (especially women), but also behind-the-scenes personnel and writers often complain of age discrimination. "I've got a refrigerator that is older than you," is one of the imaginary, but nonetheless cherished bywords, spoken *en l'esprit d'escalier* by a writer to a young producer.

In one oft-repeated story, Howard Rodman, a highly respected Hollywood writer, was asked in his sixties by a very young television executive:

"And what have you done, Mr. Howard?"

"Before or after you were born?" the writer inquired.

Not Going Gentle

Writers in television find that they begin to lose their market value after forty. Most withdraw quietly, as writers tend to do, drifting into other lines of work or living off residuals. But one-time columnist Burt Prelutsky, who had written for such shows as M*A*S*H, recently went public with the problem:

Having been in the business for twenty years, and having won various awards writing some of the top series, I thought I'd covered the bases pretty well: I hadn't locked myself into any one genre; I had done both two-hour movies-of-the-week and half-hour comedies. I figured no matter what happened in TV, I'd have *some* work, but it just dried up. After I had landed only one job in two years, and I lost my Writers' Guild health insurance, I decided on drastic action. I took out a full-page ad in *Daily Variety*. Though it was very risky, both for the cost and for the impression it would make, I knew there were good precedents: Bette Davis had advertised her availability, and so had Oscar Hammerstein back in 1939, after a series of flops. I tried to make my copy tongue-in-cheek, listing all the usual things I might have done to deal with my condition. I pointed out that I had not yet hired my neighbor's twelve-year-old and tried to pass him off as my writing partner; I had not bought a toupee yet—and other stuff like that. The basic message was that I wanted and needed a job.

So what happened? Mainly, a lot of other older, unemployed writers called me to say "good for you," congratulating me on my courage. I don't know if it took a lot of courage; I didn't have many options. I did get a call from Steve Cannell, but I couldn't come up with the right story line that worked for him, and I did a rewrite of a *Star Trek* episode—they are all aging on that show. I was also asked to write a piece for weekly *Variety* to explain why I had taken out the ad in *Daily Variety*. They even ran my ad as an illustration, in smaller size but for free. But the punchline came when an actor sent his picture and résumé—wasting the stuff on me, who had just advertised that I couldn't get a job! I guess he wanted me to know that he was in worse shape than I was. It reminded me of the old story about the Polish actress who arrived in Hollywood and she was so dumb that she went to bed with a writer.

Too Many Indians

When George Schlatter was fired as producer of *The Judy Garland Show* in 1963, he took it in fairly good humor. He managed to dig up an Indian warbonnet and burst into Judy Garland's dressing room at CBS, yelling: "I refuse to be fired!"

Schlatter was echoing how Garland herself had been let go from MGM during the filming of *Annie Get Your Gun*. Dressed and made up as an Indian, she got the news that the studio—her home and workplace for more than half her life—was tearing up her contract. "You can't fire an Indian," she was reported as saying, "I don't even have a reservation!"

George Schlatter was dismissed in the middle of preparing the sixth episode of the show, which was abruptly cancelled. Guest stars Nat King Cole and Jack Carter were trying to figure out why CBS would have taken such a drastic step.

"They cancelled the show," Cole speculated, "because I'm black."

"No," Jack Carter said, "they cancelled the show because I'm Jewish."

"No," George Schlatter stepped in to take credit, "they cancelled it because I'm difficult!"

An Offer Not to Be Refused

Film director Norman Jewison stepped in to produce the rest of *The Judy Garland Show*, and George Schlatter went on with a distinguished career, which included *Rowan & Martin's Laugh-In*, one of the landmark comedy shows of the seventies. The rapid-fire one-liners, made a number of actors or their tattooed bodies famous, but the show also derived some of its shock value by revealing famous guests unexpectedly through windows or behind bushes. Such diverse personalities as John Wayne, Dick Gregory, Gore Vidal, Billy Graham, and—picture it, if you will—a hip Richard Nixon made lightning, almost subliminal appearances. Hubert Humphrey turned down the opportunity, and conservative guru William F. Buckley had to be cajoled.

"We sent him a wire," Dick Martin recalled to Donna McCrohan, "asking him to be on the show, but he was insulted that we had asked. So we wrote him back a wire saying, 'Will you appear on the show if we fly you out on a plane with two right wings?'"

Buckley couldn't resist that offer and turned out to be a great guest.

Not Everybody Who's Somebody Gets a Peabody

Jimmy Durante received in 1951 the George Foster Peabody award, the most prestigious prize in broadcasting. Since the bulk of the awards is given in serious, non-fiction categories, the comedian sensed himself out of place.

"I rubbed elbows with scientists and educators," Durante marveled after attending the award ceremonies: "Dey wouldn't shake hands wid' me. So I rubbed elbows wid 'em!"

Count on Me

Jimmy Durante's habit of malapropisms was as prominent a part of his persona as his famous proboscis. At about the time his radio gag writers came up with "It's a catastastrope!" his partner Garry Moore warned Durante:

"You know, Jimmy, if you don't mispronounce this, we won't get a laugh."

"Yeah, yeah," said the comic. "Educate me and we'll both be out of a job."

The Great One

Jackie Gleason, who cast a giant shadow over early television, was christened "The Great One" by someone familiar with both the physical and abstract dimensions of greatness: Orson Welles. The two drinking companions were at New York's Stork Club and Welles was testing Gleason on his knowledge of Shakespeare, by naming a character and asking for a speech. But then Welles posed a trick question and Gleason said:

"Wait a minute, pal, that isn't Shakespeare, that's Aeschylus," and nonetheless proceeded to give the right speech. It was at that point, as the comic related it to television critic Tom Shales, that Welles said unto him:

"You're The Great One."

Jackie Gleason once performed Hamlet's soliloquy, "To be or not to be," in his television persona of Reggie Van Gleason III. The comedian received a telegram from actor Richard Burton afterward:

"Dear Jackie, that's the first time I ever understood it."

Solitude

Garry Moore was making a guest appearance on *The George Gobel Show* on NBC. Just before the show, he passed Gobel's dressing room. The comic invited him in and proffered a swig from his bottle of whisky.

"Thanks," said Moore, "but I don't believe in drinking before the show. I'll be happy to join you for a drink afterward."

"Do you mean to say, Garry," Gobel looked at him with utter disbelief, "you go out there all alone?"

(George Burns in his book, *All My Best Friends*, tells this story with Pat Buttram being offered the drink by George Gobel.)

How to Audition

Nervous performers sometimes make little speeches to explain why they are not in their best form for their audition, or just to postpone the evil moment. Herb Sanford recalls comedian Don Adams, later the star of the spy-spoof series *Get Smart*, doing an unusual audition for *The Garry Moore Show*.

"Mr. Moore," the actor began on the studio stage, "Don Adams is not really my name. My father was a famous star in the theatre. I don't want to trade on his name . . . but if I were to tell you, a tear would come to your eye, there'd be a lump in your throat. You know him—I know you loved him—oh, well, I might as well tell you his name . . ."

There was a long pause, while Adams's face went blank. "Funny, I was talking to him just the other day. His name is . . ." The actor began searching inside his jacket pocket. "I know I had it here somewhere . . . it's . . ."

"Thank you, Mr. Adams, that's all we need," Garry Moore said over the intercom.

Adams looked destroyed.

"But, Mr. Moore, you haven't heard my act."

"We don't need to hear the rest," Moore replied, by this time in the studio. "With a start like that it's got to be great."

Breaking In

Betty White, one of the most enduring and beloved comediennes on American television, performed on an experimental broadcast as far back as 1939, and she has been on television continuously since 1949. Asked when she began in comedy, White replies:

"The first breakfast I ever had with my mother and father. My father was a traveling salesman who used to bring home jokes. Some of them were risqué, so he would tell me: 'Honey, don't take that joke to school.'"

Although Betty White grew up in Hollywood and witnessed the birth of the television industry, she faced some of the same difficulties that other beginners still have in breaking into the profession. She had to get a paying job in order to get into the union; on the other hand, most of those jobs were open only to union members. Finally, a friendly producer decided to take a chance with her, putting White into a commercial where she only had one word to say: *Parkay.* "He said I couldn't do too much damage to him with one word."

The cost of joining the Screen Actors Guild far exceeded the pay from that one commercial, and Betty White turned to her father:

"Could I borrow the initiation fee?"

"You sure can, honey," her dad said, but hearing how much his daughter would be paid for her first professional job, he added: "And if you don't work too often, we could almost afford it."

It's a Struggle

Carol Burnett got her early break in television on *The Garry Moore Show.* As her audition piece, she played the part of a shy, inexperienced girl doing an audition.

Barely into her song, Ken Welch, who was accompanying her on the piano, launched into a long series of elaborate arpeggios. Carol Burnett fell silent and watched him transfixed. After he finished the passage, she did not pick up the song, but now completely herself, strode up to the pianist and said in that funny, but completely sincere way of hers:

"O-o-oh, that's won-der-ful." She was booked on the show that week.

Although she got a few other shots on television as a result of *The Garry Moore Show*, Carol Burnett continued to collect—as do most struggling performers—unemployment checks from time to time. One time the actress filled out the usual forms at the unemployment office, listing her latest earnings of $750 she received after her first appearance on *The Ed Sullivan Show.* The clerk knew her well by then, and thought she had made a mistake.

"Don't you mean seventy-five dollars, Miss Burnett?"

"No," she replied with grim triumph. "I mean seven hundred and fifty dollars."

Gagged Man

Danny Thomas began in show business by playing beer gardens in Detroit under the name of Jacob Amos. He wanted to be a radio actor, but when he applied to station WXYZ the only opportunity was in the sound-effects department. Thomas was handed a pair of coconut shells and made his debut as the horse's hoofs on *The Lone Ranger.*

Didi Conn was playing the receptionist in *The Practice*, a sitcom starring Danny Thomas. One of the running gags was the terrible coffee she made each morning for the doctor, who would then gag in more ways than one. It also afforded opportunities for the propman for visual gags, like making the spoon stand up. For one show Conn mixed some hot tabasco sauce into the coffee, which exploded in Danny Thomas's throat.

"What are you trying to do?" he hissed at her. "Or d'you think I can't act?"

Birth of a Salesman

When producer Norman Lear first came to Hollywood after World War II, he tried many things, including selling baby pictures door to door. Ed Simmons, a fellow salesman with whom he worked the territory, had ambitions as a comedy writer, and the two of them started having four-hour lunches. From these were born some skits, and one of them finally sold for $25.

As his skills improved and ambition grew, Lear wanted to break into the big time but did not know anyone in show business. One day he and his partner came up with an idea that they thought would work for Danny Thomas. Lear called up the William Morris Agency and said he was Merle Robinson (the actual name of a childhood friend) doing a story for the *New York Times* and needed to reach Danny Thomas immediately. He sounded so authoritative and harried that the receptionist gave him a number. When Lear tried it, Danny Thomas answered.

"You don't know me," said the young writer, "but I have a fantastic piece of material for you."

"How did you get my number?" the comic asked. Lear evaded an answer, but to his amazement, Thomas then told him that he was in the middle of working on a routine with his accompanist for a Frolic at the Friars Club, where unfortunately everybody knew all his material already. He needed help in a big way.

"I have never done this before," Thomas explained, "but how fast can you be here?"

Lear swallowed hard, remembering that he had only an idea, nothing on paper.

"It will take a couple of hours."

The two fledgling writers spent the time writing down some notes and then went to see Danny Thomas, who loved the sketch and bought it. It was a hit at the Friars' Frolic, where an enterprising agent heard it and called the next morning.

Norman Lear was launched on a career in television.

Recycling

At age eleven, Norman Lear, the creator of many of the most successful situation comedies on American television, won second prize in a poetry contest with a poem called "City Lights" which was published by the New York *Daily Mirror*.

His father was impressed enough to commission the young poet to write something that would accompany a gift to his wife on their next anniversary. The anonymous poem was framed and hung permanently in the Lear household. Meanwhile, Norman turned into a writer, and thirty years later, when he felt the opening episode of *All in the Family* required a sentimental poem, he remembered those lines he composed at the age of eleven ("All the years I have been with you, / I have shed many a tear . . ."), and decided to steal them back.

The Killer Bee

One of the early guests to host *Saturday Night Live* was Rob Reiner, then starring in Norman Lear's popular sitcom *All in the Family*. Reiner's demands and star turns did not make him popular with the regular members of the cast. The Not Ready for Prime Time Players were newly enamoured with a running gag where they dressed up as bees. Reiner did not like the gag and had said so in rehearsal. He was annoyed when during the live performance John Belushi appeared in his bee outfit playing a waiter in a restaurant sketch. He began serving Rob Reiner and Penny Marshall.

"I was told no 'bees' when I signed up," Reiner protested, breaking out of his role. "They're not helping . . . they're ruining the show."

Belushi rose to the unexpected interruption, and incorporated the mounting frustration the whole company had felt with Reiner during the week:

"I'm sorry if you think we're ruining *your* show, Mr. Reiner," he mocked while buzzing around the guest's head. "But see, you don't understand—we didn't ask to be Bees. You see, you've got Norman Lear and a first-rate writing staff on *All in the Family*. But this is all they came up with for us. Do you think we like this? No, no, Mr. Reiner, we don't have any choice."

Egged on by audience applause, Belushi worked himself into more of a rage: "You see, we're just like you were five years ago, Mr. Hollywood California Number One Show Big Shot! . . . We're just a bunch of actors looking for a break, that's all! What do you want from us! Mr. Rob Reiner! Mr. Star! What did you expect? *The Sting?*"

Baba Wawa

When Barbara Walters left NBC for ABC in the summer of 1976, her five-year five-million-dollar contract made her the highest paid journalist in the world. The highly publicized deal reaped a whirlwind of controversy, especially since Walters's career from being a writer and researcher on *The Today Show* had never included any journalism. Her rapid rise in television was all the more remarkable given her well-known speech impediment of substituting *w* for *r*, which quickly became the butt of cruel jokes and intense satire. The good people of *Saturday Night Live* on her old network immediately created the character of Baba Wawa, which made both the anchorwoman and her impersonator, the late Gilda Radner, famous. Walters took all the kidding in good humor, and when she met Radner at a diplomatic reception, she asked the actress to do the spoof for her. Later she told the *New York Post*: "It's a good thing I'm not teamed with Tom Brokaw. He can't pronounce *l*'s."

Although Walters worked hard on her diction, some words continued to defeat her. Once, when Mount Ararat had to be mentioned in the news, inevitably it came out as Mount Awawat. As a joke, she asked the newswriter why he couldn't change the location to Mount Kisco.

Can We Borrow You?

Karen Salkin, a comedy actress who was making her own programs for community cable, was excited one day to get a call from one of the producers of *SCTV*, the Canadian comedy show, which became hot in the early eighties. The producer was very flattering about her cable show, and asked for some of the tapes, saying that *SCTV* might need a character just like her. The actress, thrilled by the prospect, said she would love to have a shot at it, and that *SCTV* was one of her favorite shows, and—

But the producer cut Salkin off, since that was not exactly what he had in mind. He and his colleagues merely wanted to study the tapes and then have one of the *SCTV* regulars play Karen Salkin. Neither her agent nor manager found the request surprising. Perhaps the only unusual aspect of wanting to steal her character was that she was asked.

The Birth of Python

Monty Python's Flying Circus, one of the most successful and bizarre comedy shows on the BBC, or any television for that matter, got its name after several brainstorming sessions. Some in the group wanted to call it *Owl-Stretching Time*, while actor John Cleese suggested *Bunn, Wackett, Buzzard, Stubble and Boot*, an imaginary football team's forward line. Barry Took, one of the writers, had been rechristened Baron von Took by Michael Mills, head of comedy at the BBC, and from there it was but a step to Baron von Richthofen, the celebrated German flying ace of World War I. Another of the actors, Michael Palin, "happened to spot the name Gwen Dibley in a local newspaper," wrote George Perry in his chronicle of the Pythons, "and for a while the show was to be *Gwen Dibley's Flying Circus*. Then out of discussion the name Monty, representing a shady sort of theatrical booker, the sort of man who fixes a fourth-rate act up with a week at Workington, and Python as an unlikely surname, were gloriously combined, producing an octosyllabic title of pleasant resonance. It also put paid to a proposal to give the show a new title each week, an idea that the BBC had viewed with horror and disbelief."

The catchphrase "And now for something completely different," popularized in the early Monty Python episodes, came from the fact that the show originally replaced a Sunday night repeat of a dull, religious program. "Something completely different," as any fan of Python would testify, was simply truth in advertising.

No Cast Unstoned

Graham Chapman, one of the Monty Python gang, appeared as himself on a talk show, during which host George Melly discussed his homosexuality. A viewer wrote in, complaining that the actor had not revealed his identity—which he had—and she enclosed some prayers for his soul, with a biblical quotation: "If a man lie with another he shall be taken out and killed."

The letter was forwarded to the Monty Python production office, where fellow Python, Eric Idle, composed a reply, assuring the correspondent:

"We have found out who it is and we've taken him out and killed him!"

The greatest controversy stirred by the Pythons was their film *The Life of Brian*, a parody of biblical epics about the life of Christ. John Cleese and Michael Palin tried to defend themselves on a BBC talk show against merciless attacks led by Malcolm Muggeridge and the Bishop of Southwark, who acidly assured the actors that they would be receiving thirty pieces of silver for their work. The film was banned in various local authorities in many parts of Britain, including one district that had no cinema within its boundaries.

Collaborators

Frank Muir and Denis Norden, who have amused BBC audiences for decades with their eccentricities, word-plays, puns, and vast store of curious knowledge, met after World War II through Ted Kavanagh, the creator of *ITMA*. After auditioning for him, they went their separate ways but that very afternoon they both happened to go to the same cinema on Regent Street. During a rather ordinary piece of dialogue in one of the films on the double feature, two laughs rang out in the auditorium. Muir and Norden looked around and that is how they found each other again—through their peculiar funny bones.

Denis Norden got tired of being asked so often how the collaboration with Frank Muir worked.

"We do actually and physically write together," he once explained. "You see, we've got a very, very big pencil."

The I and I

Television writing involves offering broth to a great variety of would-be cooks, sometimes even without the expertise how to spoil it properly. Bill Idelson, writer of such famous shows as The Odd Couple, *recalls some problems in creating a series from* The King and I, *a show that never made it. We also get a glimpse into the life of a Hollywood writer.*

If I felt relief at getting away from the stressful *Odd Couple*, it was short-lived. *Anna and the King* (the title of the TV version) had its own tribulations. One was Yul [Brynner], who proved to be a vain, pompous, whimsical pain in the ass. Another was his co-star, Samantha Egger, who could be spoken to only through Yul, another was Gene Reynolds, my executive producer, who was often irrational and had a Napoleonic complex. But the biggest problem was stories. It was a period piece, set in a place no one knew much about. It's difficult to get good comedy out of Siam in 1860. Especially from Hollywood comedy writers who knew Brooklyn and Manhattan and the Brown Derby and not much else.

I went through the same sort of shotgun search for writers that we had on *Love*. There must be someone who could write Siam in 1860, and this peculiar, exotic world of the King. Again, a writer came through the door who looked like he could save our ass. This time it was a guy I knew pretty well, Jerry Mayer, one of the working comedy writers in town. He had a good story: A man-eating tiger is reported in the area, and the King decides to go out and hunt it down. He will take his number one son, the Prince who will succeed him, along on the hunt. The young Prince is so terrified at the prospect that he takes ill. The King is bewildered by this and Anna must intercede so that King has some empathy for the boy.

It was a damn good script, exciting, suspenseful, with emotional values and it went well. However Yul was not satisfied. "This writer—Mayer—does not understand the King. Don't use him again." Jesus, we'd found the one writer in the whole town who could write the damn show, and Yul forbids us to use him again! What an asshole!

Also when it came time to shoot the show, it screamed for the body of the tiger to be brought in on a pole, when the King made his triumphant return to the palace. Gene Reynolds balked at this. "Tigers are an endangered species," he told me. "We're not going to show one dead on TV."

"But Gene," I pleaded. "We've got to have the tiger. We've been talking about it all during the show. Besides they weren't endangered in 1860."

"No tiger."

"I'm not talking about killing a tiger. I'm talking about getting an old tiger skin and stuffing it."

"No."

"Listen, how about the tiger skin. Just the skin?"

"No, I'm against the wearing of wild animal furs."

"How about a *fake* skin?"

"No, that stimulates people to buy wild animal furs."

"Gene, this is 1860."

"It doesn't matter. What we know now, should color our perception of the past."

"What does that mean? You wanna rewrite history?"

"No tiger. No tiger skin." His assistant, Burt Metcalf shrugged and rolled his eyes at me, unseen by Gene. Then they both got up and left my office.

I should have gone over Gene's head, but I was determined to play it straight, definitely a mistake when dealing with the likes of my executive producer. We shot the scene without the tiger or his skin and it came off O.K. but that kind of thinking made our show one of the endangered species.

Screw Yul, we hired Mayer to do another script. It was a dandy, better than the first. The scripts came out of mimeo and were distributed. Gene called me in a panic. "Did that script go to Yul with Jerry's name on it?"

"I don't know. Maybe. Probably."

"God, see if it's in his dressing room. He may not have seen it yet—he's rehearsing. Go!"

I ran down to the stage, stole by Yul as he was going through his scene with Samantha. He waved at me, and I waved back and wiped my forehead with my handkerchief. I stole into his dressing-room. Yes, there was the script! Feverishly, I tore out the title page and crumpled it into my pocket.

We read through the next week's script after shooting that day, seated around a long table. When we'd finished, Yul said, "Now, this is a very good script."

"Well, Yul," I said, "now I can tell you. This was written by Jerry Mayer."

"But he still doesn't have the King right," said Yul, as if he hadn't heard me. "Don't ever hire this guy again!"

Yul was the old-fashioned, classic Prima Donna. He insisted on being picked up at his home in Malibu every morning at five-thirty. He arrived on the set not knowing line one. He learned his lines while shooting, wasting thousands of dollars of film. Just because he was working at a movie studio—Fox—which he always called "Eighteenth Century Fox" he felt he could do this TV show like a big-time movie, where they shot maybe a few lines a day, instead of many pages. He never understood this was for TV, and in spite of the ratings, which were poor, he had an absolute conviction that the show would be a big hit, because he was in it.

When the handwriting was on the wall, and everyone in Hollywood could read it, I got a call from Grant Tinker at MTM. "We need a producer on *Bob Newhart*, he told me, and I gather you'll soon need a job."

All That Glitters

One of the many tabloids that live off backstage intrigues about the public's favorite television series, reported once on a work stoppage on the set of the popular sitcom *The Golden Girls*. It was during the first full rehearsal before an audience, according to the story, that the four actresses in the title part became so upset by the script that, calling it garbage, they stormed off the set. The anonymous source claimed that the writers scurried back to their offices in panic and within twenty minutes came up with a satisfactory script, which the ladies read and then resumed the rehearsal.

Not one word was true about the story, Betty White recalled, but the writers laminated and framed the article just to remind themselves that if they could write a good comedy episode in twenty minutes then they were much better than they had been credited, and to show management that they were worth every penny of their very high salaries.

Genie

Just Our Luck, a sitcom that ABC tried out briefly in 1983, revolved around a genie enslaved to do his master's bidding. Since the genie was played by a black, and the master by a white actor, not surprisingly the show drew protest from groups such as the National Association for the Advancement of Colored People and the Black Anti-Defamation Coalition. Commenting on the genie's promise to serve his master "for two thousand years or until your death, whichever comes first," the Coalition countered: "We have already put in 400 years. Hopefully death for this show will come first."

The Death of Comedy

Larry King was hosting Jackie Gleason on his show, and they were discussing the state of comedy. King asked what had gone wrong with Jerry Lewis's career.

"Jerry Lewis found taste," said the Great One.

Intimations of Mortality

Georgie Jessel in his latter decades became the best-known toastmaster in America. He was also much in demand for eulogies at funerals. After he wrote out his last instructions, the dying Ed Wynn is said to have summoned his son, Keenan, and said he had left one thing out.

"Promise me," the comic insisted, "no Jessel."

Jessel became so well known for his funeral orations that once when he was about to introduce Harry Truman, the president asked him: "Georgie, do me a favor. Don't get too sentimental about me. Every time you do, I think I'm dead."

Bob Hope was visiting the deathbed of Barney Dean, a comedy writer who had once worked for him. According to George Burns, the dying man motioned Hope to come closer and then whispered:

"Anything you want me to tell Jolson?"

Groucho Marx had one final request: "Bury me next to a straight man."

Take Note

The inscription on the tombstone of Ernie Kovacs at Forest Lawn reads:

Ernie Kovacs
1919–1962
Nothing in Moderation

CHAPTER 10

Just Kidding!

The Joke Is Wild

Ernie Kovacs, the great comic of early television, was fond of practical jokes, especially on the air. Despairing whether anybody was watching one of his early shows, Kovacs suggested that people call Tony De Simone to wake him up. He put the musician's phone number on the screen and then forgot all about it, until he heard not from De Simone but the telephone company, asking him never to do that again.

Edie Adams also tells the story in her recent autobiography how she and Ernie Kovacs were traveling once on the train between Philadelphia and New York, when it screeched to a sudden stop. A great many people got bruised, and the railroad's insurance company arranged to interview them. Adams and Kovacs were short of material for that week's live television show, so they deliberately made an appointment for the insurance man to come during the show. The man was embarrassed enough to be grilled by Ernie Kovacs about the incident, but the real horror came when Edie Adams was wheeled on the set, bandaged up to her ears. After allowing the audience to roar with laughter at the assessor, Edie Adams finally took pity on him. She leaped out of the wheelchair, and said:

"Nothing's wrong, but I ripped a pair of stockings." The man happily scrambled off the set and bought her several pairs.

Soft Touch

The critic Alexander Woollcott, a man of great physical bulk and uncertain sexuality, had a popular radio program called *The Town Crier*. Under his acerbic exterior, Woollcott was known in his circle, which included the famous Algonquin Round Table of wits, as a sentimentalist. Every actor or playwright in need of money or introduction knew that he was the softest touch on and off Broadway. These qualities made Woollcott a perfect target for one of the classic hoaxes in broadcasting.

One day *The Town Crier* office received a typed letter, purporting to be from two elderly maiden sisters living in great poverty in Troy, upstate New York. Their only comfort was listening to "the greatest person on American radio." Woollcott, who was enormously vain, believed them. Many letters followed. Then one of the spinsters, signing herself Susan Lovice Staples, wrote that her sister Minnie was desperately ill, and that before she died, it would give her great comfort to hear Woollcott recite the Twenty-third Psalm. Woollcott naturally obliged and read David's song to the accompaniment of the full CBS orchestra. It brought an explosion of public response, all wanting to help the poor, lonely sisters.

A letter arrived a few days later, describing how Minnie had passed away with a smile on her face, listening to her favorite personality reading her favorite Psalm. Woollcott was touched to the core and wanted to help Susan. But none of the letters had a return address, and Woollcott made several futile appeals over the air for information. There was now silence, until one day a note came, telling of Susan's death. It was signed simply as Nurse O'Brien, and described how the old lady "stretched out both her arms, like she was taking hold of your hands. . . . Once she called your name and blessed you and once she said something about some still water."

By this time obsessed, Woollcott dispatched researchers to Troy to find medical records, a death certificate, any clues as to the identity of his extinguished fans. Months of search turned up nothing, and Woollcott was forced to admit that he had been hoaxed. He accused various friends, and his prime suspect remained Charlie MacArthur, the most famous prankster in his circle. But the playwright would likely have claimed credit, and he emphatically denied having masterminded the unsolved hoax.

Wherefore Art Thou Romeo?

Woollcott was an incorrigible ham (he toured the country in the Kaufman and Hart comedy, *The Man Who Came to Dinner,* playing the title character based on him), and he invited Charlie MacArthur's famous wife, Helen Hayes, to come on *The Town Crier* program to do the balcony scene in *Romeo and Juliet.* Thanks to the unseen magic of radio, the audience did not know that Miss Hayes was almost nine months' pregnant, and that the rotund Woollcott made the unlikeliest of Romeos.

"I looked like a watermelon," Hayes recalled, "and Woollcott having a pretty big stomach himself—this Romeo and Juliet could never possibly have reached within an arm's length of each other. But there we were doing it. He was an absurd Romeo. Aleck just liked to read the lines. As the Nurse, of course, everyone agreed he was much better."

Chalk One Up for the Dummy

Alexander Woollcott, who had helped out the young Orson Welles at the beginning of his career, sent him the following telegram after he heard of the credulity of some who had been listening to *The War of the Worlds* broadcast, instead of Edgar Bergen on a rival network:

DEAR ORSON. THIS ONLY GOES TO PROVE MY CONTENTION THAT ALL INTELLIGENT PEOPLE LISTEN TO CHARLIE MCCARTHY. ALEXANDER WOOLLCOTT.

Mars Calling

The first Martian sensation was not caused by Orson Welles but came about accidentally in August 1924, when WHAS in Louisville, Kentucky, tried to do a remote broadcast of army maneuvers near Fort Knox. In his Microphone Memoirs, *station manager Credo Fitch Harris describes the widespread impact of the broadcast.*

The zero hour approached. Standing with me by the transmitter our technician was even more nervous than the captured city. For a remote pick-up from such a distance had never been attempted so far as we knew and, moreover, we were now becoming alarmed over what the blasts of cannon fire might do to our tubes. Tubes were not designed for such unexpectedly harsh treatment. Would the thunderous explosions frizzle them up? Would they burst? Melt? Or take it?

Staged for the first great clash between Blue and Red armies, when artillery, machine guns and rifles would be heavily engaged, General Tyndall thoughtfully ordered Colonel Hamer to a position in the very thick of it. This was close between two 3-inch field pieces, one of which had orders to fire on the minute and half minute, the other on the quarter after and the quarter of. Thus we would get nearby bursts every fifteen seconds, and the intermediate spaces were to be filled with incessant machine gun and rifle fire.

I opened the station, saying that Colonel Hamer, at the front, would report the battle up to this point—which he did in a splendidly dramatic and understandable way. Then the firing began.

The technician and I looked at each other. That first 3-inch gun, far from damaging the tubes or splitting our eardrums, merely sounded as if someone had struck a dilapidated, partially cracked old gong.

. . . Unsuspected by any of us, the old-fashioned microphones of that day could not pass through their little insides those frightful blasts, because the granulated particles of carbon used in their construction instantly froze beneath the sound impacts, feeding us a distorted, imperfect signal. As instantly, it is true, they unfroze, permitting the milder "zips" to sound quite natural.

Colonel Hamer, letting us have a run of this for about twenty minutes, took up his story of the battle again. After that we once more plunged into the midst of shell fire, and for another twenty minutes the regularly spaced bongs! zip-zip-zip-z-z-z-zip-zip-zips held a large population enthralled.

The fighting was fast and furious. Would our gallant Blues destroy that imaginary horde of ruthless invaders? Colonel Hamer stood ready, should Fortune turn her face from us, to give immediate warning, so that all in the enemy's line of approach might hastily evacuate their homes.

A Chicago radio weekly, under a caption extending across its first page, carried a great story commending us for that sensational broadcast. But, while the battle raged, we had no suspicion of wider sensations that were being stirred up, nor what strange story distant night editors from coast to coast were excitedly reading, hot off the news wires.

Now it so transpired that that very Friday our Earth and Mars were nearer to each other than they had been for several years. Weeks beforehand astronomers had faithfully polished their lenses. Also for weeks imaginative writers had set the country wondering if the miraculous infant, radio, might not be able to gather unto itself signals from our neighbor planet. Proponents of the theory that those mystifying "canals" were dug with the distinct purpose of attracting the attention of us Earth dwellers, began writing to inquire when and how we intended broadcasting messages to the Martians. Several had suggestions. Some wanted to come to the station and help. A few told us what to say—especially one note from a little girl. I quote it:

"I wish you'd ask them if they have dogs there, and if there are any Scotties. My Scotty looks so long at the moon sometimes I wonder if he is seeing other

Scotties, too. I asked Mother and she said maybe, but I don't think she was listening."

At any rate, the impetus with which such letters increased warned me of another avalanche of madcap mail. But, most assuredly, I was not prepared for what did come!

The first thunderclap appeared in a news dispatch to the *Courier-Journal* from Boston and told a weird tale of a radio specialist who was tuning his receiver at random when suddenly strange signals started coming in to him. They resembled the ringing of a gong and ended in abrupt z-z-zips. He dashed to the telephone. Signals from Mars! There was a lot more, of course.

A few hours later came another sizzling hot story from Vancouver, British Columbia, also telling the world that mysterious signals had been picked up at that point, which were causing our Canadian friends to wonder if the planet Mars were trying to establish communication with us.

No known code, International, Morse or private, contained anything resembling them. They started on a low note, and ended with zips—quite a number of zips. Then the low note came again, and more zips followed. Radio people in that corner of the Dominion declared they had never heard the like of it before. Aside from Mars trying to call us, other explanations just did not exist! And one listener firmly supported this hypothesis by the precise timing which spaced each of the gongs, which seemed conclusive evidence that some intelligible communication agency was at work somewhere out in our constellation.

It is useless to mention other clippings that found their way to us, but most of them showed a self-indulgent prodigality with bold, black types.

From Boston to Vancouver, and sweeping into the South, numerous editors played up those "signals from Mars" with splendid éclat—until the true explanation hit the wires when, of course, everything went suddenly quiet.

Panic in the Streets

Though Orson Welles's *War of the Worlds* broadcast in 1938 is the most famous of hoaxes perpetrated on a radio audience, it was by no means the first of its kind. On the evening of January 16, 1926, the BBC interrupted a religious lecture being broadcast from Edinburgh to bring live reports of disturbing events taking place around the heart of London. Among other things, an eyewitness claimed that "the Houses of Parliament are being demolished by an angry mob equipped with trench mortars. The clock tower 320 feet in height has just fallen to the ground, together with the famous clock, Big Ben, which used to strike the hours on a ball weighing nine tons. One moment please . . . Fresh reports announce that the crowd has secured the person of Mr. Wurtherspoon, the minister of traffic, who was attempting to make his escape in disguise. He has now been hanged from a lamp post in Vauxhall. London calling . . . That noise you heard just now was the Savoy Hotel being blown up by the crowd . . ."

Despite its generally broad humor, the broadcast was taken in earnest by hundreds of Britons, who flooded the telephone exchanges and telegraph offices, even after the government and the BBC had issued official explanations. Many people had missed the introduction to the program by the Reverend Ronald A. Knox, who intimated that a burlesque would follow. They then took seriously a prototypical piece of Monty Python humor that unemployed workers were rioting in Trafalgar Square, "led by Mr. Popplebury, secretary of the National Movement for Abolishing Theatre Queues."

Martian Chronicles

Although audiences were again warned, the Mercury Theatre's adaptation of H. G. Wells's science-fiction nightmare about a Martian invasion caused much greater panic than the BBC's broadcast twelve years earlier.

Thousands of New York and New Jersey residents fled from their homes; hospitals treated hundreds of patients for shock. Some of the more adventurous citizens took up arms and went searching for the octopus-like invaders somewhere near Princeton.

In a wide area, radio, newspaper, and police headquarters were swamped with inquiries. Church services, where the devout went to pray for spiritual aid and comfort, were disrupted by others spreading alarm. The staff of the Memphis *Press-Scimitar* was called back to work on an extra edition.

Thousands of miles away at a southwestern college, "the girls in the sorority houses and dormitories huddled around their radios trembling and weeping in each other's arms. They separated themselves from their friends only to take their turn at the telephones to make long distance calls to their parents, saying goodbye for what they thought might be the last time. This horror was shared by older and more experienced people—instructors and supervisors in the university."

Hadley Cantril of the Office of Public Opinion Research at Princeton, who first published his study of *The Invasion of Mars* two years after the broadcast, cites such minor victims of the hoax as George Bates, an unskilled laborer in Massachussets, who wrote him: "I thought the best thing to do was go away, so I took $3.25 out of my savings and bought a ticket. After I had gone 60 miles I heard it was a play. Now I don't have any money left for the shoes I was saving up for. Would you please have someone send me a pair of black shoes, size 9-B."

There have been several attempts to explain how so many people could have been taken in. A great deal had to do with the skill of the broadcast itself, using specific place names, routes, and streets. Having experts and scientists on the program, albeit impersonated by Welles and others, also helped to suspend the listener's disbelief. The listening habits of the audience also came into play. Many often missed out on introductory warnings or comments before a program. Some had tuned in first to Edgar Bergen's competing program, and came in late.

The times were grim, too. *The War of the Worlds* was broadcast a few weeks after the annexation of Czechoslovakia by Adolf Hitler. Some people listening confused the Martians with a German invasion. A parallel may have been intentional. The journalist Dorothy Thompson, who called the broadcast "the story of the century," explained that "Mr. Orson Welles and his theatre have made a greater contribution to an understanding of Hitlerism, Mussoliniism, Stalinism, anti-Semitism and all other terrorisms of our times than all the words about them that have been written by reasonable men."

Finally, some believed afterward that because of such historical circumstances, a similar broadcast could not cause like panic ever again. Hadley Cantril of Princeton came to the opposite conclusion; he wrote in 1966 that "of course it could happen again today and even on a much more extensive scale."

Indeed, many years after its original broadcast, the Orson Welles version of *War of the Worlds* was translated into Spanish. In one of the Latin American countries where it was being played, the reaction changed from apocalyptic hysteria to blind fury when listeners found out that the whole broadcast was a hoax. According to news reports the angry mob in one city burned down the radio station, killing several actors.

Horsin' Orson

Orson Welles enjoyed playing practical jokes of a less momentous kind. He once scared Charles Martin, a radio drama producer, nearly out of his wits by stumbling on stage and spilling his script in a dozen different directions.

With the sign ON THE AIR already flashing, and the audience watching in disbelief the great man on all fours hopelessly trying to gather up every sheet, Martin saw no hope that the live broadcast would start on time, and was frantically trying to think up what to do—when Welles, a talented amateur magician, whipped his real script out of his pocket, and was on his feet in a flash, in front of the mike, ready to go, to the delight and applause of the studio audience.

On another occasion, the great ham was a guest with Claudette Colbert on a morning program hosted by the socialite Elsa Maxwell. To underscore the unfashionable time of day, Maxwell broadcast the live radio show from her boudoir, as if her guests were attending a levee at court.

When the engineer gave his warning, "You're on the air," Elsa Maxwell introduced her guests and then turned to Welles.

"We are very glad to have you with us, Orson dear," she said graciously, turning over the mike to him.

"Is that so?" Welles thundered into the instrument. "Well, phooey! Who wants to get up this early on this lousy radio program? For what?"

The blood drained from Elsa Maxwell's face, and Colbert looked like she was about to be sick. But Welles carried on, improvising upon his indignation for the next couple of minutes. Finally "Horsin' Orson"—as columnist Earl Wilson called him—let his half-dead host know that he was merely kidding, having set up the hoax with the engineer and the announcer, to start some minutes before the actual show.

Big Guns

Just a short time after *The War of the Worlds* broadcast in America, London went through a smaller trauma with the production of a very early television play in November 1938. Royston Morley decided to use real howitzers for his production of a wartime drama, *The White Chateau*. The producer arranged for the local Territorial Army to bring up a battery of the big guns to Alexandra Palace, where the television studios were located.

The brief shelling shattered some windows and, according to newspaper accounts, frightened old people and children "out of their lives."

The broadcast took place during the jittery times of the Munich crisis, and people thought that war had broken out. Some panicked and rushed out into the gardens where they began digging shelters. Royston Morley was riding a bus to work, when he saw a newspaper poster: "BBC PRODUCER TERRIFIES NORTH LONDON."

During these early days of live television, repeat broadcasts had to be done live, too; in this case the BBC would not allow the use of guns again. Morley tried to use fireworks instead of the howitzers, but a large crowd had turned out from the neighborhood specifically to see the big guns. They were so disappointed that they started booing, which ruined the sound portion of the broadcast, forcing the producer to switch back to the studio for the other, less exciting scenes.

Local News

In the forties John Chancellor and Hugh Downs were working for the same NBC radio station in Chicago. Chancellor wrote the news copy that Downs read on the air. One Sunday, Downs drove through the early morning fog to the studio, and in his hurry left the headlights of his Packard on. As he read a five-minute newscast that was being fed live to the network, Chancellor tiptoed into the studio and gave him a late-breaking item.

"The owner of the 1948 Packard parked in front of the Merchandise Mart," Hugh Downs informed the nation, "is going to be awfully surprised to find his battery dead. He left his lights on."

The Gizmo

From the earliest days of American broadcasting "the talent" was expected to pitch products of the program's sponsors. In the fifties Betty White worked on a daily show called Hollywood on Television, *performing on camera for five and a half hours a day, and sometimes doing as many as fifty-eight commercials live, some of them lasting two or three minutes. To make things more complicated, her colleagues would also get into the act off-screen, with the sole purpose of making her break up. This pastime is known in the theatre as "corpsing" because it is often directed at characters who are supposed to be lying dead still on the stage. Betty White recalls an incident from those palmy days.*

I just got this new piece of copy on a marvelous brush which would be attached over the kitchen faucet, so that each time you turned on the water, it would come out already mixed with soap and ready to be applied with the brush. I was enchanted by this simple invention and therefore found it easy to pitch. For some reason, our newsman, Jimmy McNamara, thought that there was something hysterically funny about this gadget, and so did the producer, Al Jarvis. Once, when they knew that I was getting ready to pitch, the two of them got behind one of the flats where they had two large buckets of water. They kept pouring the water from one bucket into the other to and fro—you could imagine what it sounded like back there. The mike didn't pick it up, but I could hear it, of course, really well, and it was most distracting. I also knew that this was a new advertising account and that we didn't want to lose our sponsor, so I kept trying to concentrate and explain exactly how it worked. I knew the principle of the thing, but when it came to explaining how the soapy water came out through this . . . through your . . . what does it come out of . . . I was going to say whatchyamacallit, but I was looking for something that sounded more professional and technical, so I said—your gizmo . . . ! Well, I lost it. Bill Niebling was the cameraman, and when he laughed, the tears did not come rolling down, the tears shot up. I couldn't see his face, but I saw waterdrops spurting all over the set. Anyway the whole thing was a disaster, I survived because Al took the blame on himself, but the people with the marvelous gizmo somehow never came back to us.

Like No Tomorrow

Lowell Thomas broke up easily, with or without help from his colleagues. His newscasts usually ended on a lighter note, and Thomas seldom read through the final item before the broadcast. One evening he was handed a sad story about a dog that had been shipped across the country by rail. The dog escaped during the trip, and when the railway employees caught him, they slammed the cage door, cutting off his tail. The owner sued and was awarded $35,000 in damages, the announcer told the nation, "which is an awful lot of money for a tail."

Thomas broke up and could not even give his traditional sign-off: "And so long until tomorrow."

The next morning he apologized to his listeners, promising that it would not happen again. Then he read the first news item involving Sir Stafford Cripps: "Today, the British ambassador to the Soviet Union, Sir Stifford Crapps . . ." The rest of the broadcast dissolved into organ music.

According to some version of the legend, though not his own, Lowell Thomas had recovered sufficiently the next day to apologize for his spoonerism, beginning: "The name that gave me so much trouble yesterday was Sir Stifford Crapps . . ." Then he began shaking with suppressed laughter, and the program had to be cancelled a second time.

One story that Thomas tells on himself involves a news item about a woman in Flint, Michigan, formerly a circus bareback rider, who had just given birth to her twenty-second child. He thought his subconcious must have taken over, because he started to pay tribute "to this Blue Star mother made of Flint." Everybody in the studio—the announcer, the engineers and the production people—fell on the floor in convulsions. "Though every listener heard laughter," Thomas wrote later, "few actually caught the boner which was even bluer."

Don't Try This at Home

Johnny Carson began his broadcasting career on a small Omaha radio station, where he fabricated celebrity interviews by intercutting new questions to answers he spliced from old tapes. For example, he got hold of an interview with Patti Page where he heard her reply to the question, "When did you start performing?" "When I was six," said the singer, "I used to get up at church socials and do it." Young Carson then replaced the original question with: "I understand you are hitting the bottle pretty good, Patti—when did you start?"

The doctored tapes enjoyed an underground reputation.

With Malice Aforethought

One of Johnny Carson's favorites is that Master of Malice, Don Rickles. When he was guest host for one of the episodes of *The Tonight Show* sometime in the early seventies, Rickles became enraged by the fact that one of his jokes had fallen flat, and smashed a cigarette box on Johnny's desk. Upon his return, Carson noticed the damage at once. "That's an heirloom," he recalled, "I've had it for nine years."

Told how the damage was done, Carson got up in the middle of the show, and instructed a camera crew to follow him into one of the adjacent studios at NBC, where Don Rickles was in the process of taping an episode of his comedy series *CPO Sharkey.*

Carson proceeded into the middle of the shot, where he confronted Rickles with the evidence of his wanton destruction and demanded restitution. Everybody was completely dumbfounded, but Carson, who apparently did this on the spur of the moment, felt triumphant.

"I really shook him," he told Kenneth Tynan afterward, "he was speechless."

Licensed Jester

Several of the big-time entertainers enjoyed being kidded by court favorites. The best contemporary example is Don Rickles's role as Frank Sinatra's licensed jester. "The Merchant of Venom" could say freely—but jokingly—all those things about Sinatra's temper and underworld connections for which reporters might get themselves knocked down.

Bing Crosby cultivated several such people in his circle. One of them was a former boxer, "Society Kid" Hogan, who attended a Hollywood party with Bing. When the Kid tried to do a bit of singing and his voice broke, Crosby kidded him with a taunt:

"What's the matter?"

"Once during a fight," the boxer came back, "a guy socked me in the Adam's apple and hurt my chords. What's your excuse, Bing?"

Publish and Survive

Jean Shepherd, talk-show host on WOR in New York, was attacking on the air the phoniness of literary cocktail parties, where people would discuss books they had never heard of. He suggested to his listeners that they try a test: at their next party they should ask everybody they meet what they thought of a novel called *I, Libertine.*

Within a few weeks of this hoax, bookstores received a hundred thousand orders for the non-existent book. Word slowly but surely reached Publishers' Row, and Ballantine Books got in touch with Jean Shepherd. Would he care to write a book with such a title?

The broadcaster sat down and wrote *I, Libertine* over a weekend. Naturally, it was an instant bestseller, and the talk-show host went on to write other books to give people at parties something to talk about.

Radio Mime

Jim Eason, the longtime talk-show host on KGO radio in San Francisco, still savors a multiple stunt he once played on the air. He had as a guest Alexis Melteff, a prankster who likes to assume different characters. This time he was posing as M. Pierre Rohe (i.e., Pierrot), the advance man for famed French mime, Marcel Marceau.

Jim Eason asked questions in English, while Pierre answered only in French. That is, when he answered at all. Soon the show took an abstract turn, with Eason calling out captions like "Frog and lilly-pad" or "Man leaning against the wind," which his guest went on to mime. This feat was achieved simply by Eason turning off the mikes in the studio for twenty seconds during the alleged mime.

After having fun, Eason decided to spread it around. He announced that anybody could book M. Pierre Rohe while he was in town. And he gave out a number to call, which happened to be that of a rival talk-show host at KSFO.

Driving home after the interview, Jim Eason tuned his car radio to KSFO and rather enjoyed the chaos he heard.

Spaghetti Harvest

One of the most elaborate hoaxes in broadcast history was an April Fool's joke played on the BBC current affairs program Panorama, *with its rather dignified host Richard Dimbleby earnestly relating a story about the annual spaghetti harvest filmed in a Swiss-Italian spaghetti orchard. Cameraman Charles de Jaeger thought up the spoof and related to Denis Norden how it was accomplished.*

Panorama's . . . first famous spaghetti harvest came from my school-days in Austria, when a master was always saying to us, "You're so stupid you'd think spaghetti grew on trees." So it had always been in my mind to do the story and I tried for several years. It was not until I was working on *Panorama* that I got the go-ahead.

I went to the Swiss Tourist Office, who said they would help, and I flew to Lugano. It was in March when I thought the weather would be sunny with flowers out. There was a mist over the whole area.

The tourist office guy took me around all over the place; not one blossom out, no leaves out. It was now Tuesday and I could not find anything and said in desperation, "What can be done?"

Then we found this hotel in Castiglione, which had laurel trees with leaves on, tall trees. So I said, "We'll do it here. Let's go down into Lugano and get some handmade spaghetti."

We did that, put the strands of spaghetti in a big wooden platter, took that in the car and we drove back. By the time we got there, the damn things wouldn't hang up. They'd dried out. So we cooked them, tried to put them on the trees, and this time they fell off because they were so slippery.

Then this tourist guy had a brilliant idea—put the spaghetti between damp cloths. That worked and we got local girls to hang them up—about ten pounds' worth. Then we got the girls into national costume and filmed them climbing on ladders with these baskets, filling them up, and then laying them out in the sun. And we said in the script, with a guitar playing in the background, "We have this marvellous festival. The first harvest of the spaghetti."

At the end of the three-minute film Richard Dimbleby said, "Now we say goodnight on this first day of April." In spite of that hint, next morning it was surprising the number of people who hadn't twigged.

Curtain Call

Though he disliked motion pictures and resisted for a long time the filming of his plays, George Bernard Shaw appeared on television during its earliest experimental phase. In July 1937, the BBC televised Shaw's play, *How He Lied to Her Husband*, and afterward the eighty-one-year-old playwright, and inveterate ham, made an appearance on the flickering screen.

"You might not suppose it from my veteran appearance," Shaw declared about this much earlier work, "but the truth is that I am the author of that ridiculous little play you have just heard."

GBS on CBS

Shaw had an irreverent attitude toward both the media and his own celebrity status. In a famous newsreel the old curmudgeon, dressed in his usual country outfit of Norfolk tweed and knickers, stuck his tongue out and made other childish signs at the camera. Following the playwright's much-publicized trip to Soviet Russia in the thirties, Caesar Saerchinger, the CBS man in London, thought he had scored a coup in persuading the socialist gadfly to give a talk for radio. Shaw succeeded in stirring up a storm of controversy with his broadcast which began:

"Hello, America! How are all you dear old boobs who have been telling one another that I have gone dotty on Russia?"

Shaw once also remarked: "Americans adore me and will go on adoring me until I say something nice about them."

Concession

In January 1959, the first American television interview filmed with Fidel Castro, after his triumphant entry into Havana, was not with a reporter, but with Jack Paar, host of *The Tonight Show*. Paar told Castro that he had just visited the lavish mansion of Batista, which was sitting empty after the dictator's flight to Miami. The new Cuban leader offered to sell the house, complete with furnishings, for two million dollars. When Paar declined, Castro asked Paar if he might want something else.

"Well," Jack Paar replied after a brief glance at Castro's bearded visage, "I wouldn't mind the Norelco shaving concession down here when things quiet down a little bit."

Parrot Joke

Ian G. Clark, owner of the pioneer radio station in Kamloops, in the interior of British Columbia, recalls the early days when he used to walk to and from work. On one of the porches that he passed there was a parrot which had its own version of the CFJC station call, "the first station in the Interior."

"You are listening to CFJC," the parrot greeted Clark daily, "the worst station in the Interior."

The Beeb

Harry Secombe, the irrepressible Welsh comedian and one of the original Goons, took over Tony Hancock's role in *Hancock's Half Hour* when the star became ill. Kenneth Williams, who was also on the radio version of the show for many years, recalls Secombe standing behind the announcer while he announced, "This is the BBC Home Service," and screaming at the top of his voice, "And serve you right."

CHAPTER 11

Fun and Games

No Information, Please

George S. Kaufman, playwright and wit, was guest panelist on *Information Please*. During the warm-up period, held also for the benefit of the guests, Kaufman failed to reply to a single question. Clifton Fadiman, one of the regular panelists, was curious: "May I ask what you have been doing, Mr. Kaufman, for the past fifteen minutes?"

"Certainly," said Kaufman, "I've been listening to *Information Please*."

Oscar Levant became known nationally for his appearances on *Information Please*. Not only did he dazzle people with his phenomenal memory for practically every scrap of music ever written, but he made people laugh with his mordant, bitchy wit, which was quite uncommon on radio in the early forties.

The columnist Franklin P. Adams said once that if he or John Kieran (another of the regular panelists) did not know the answer to the question, they simply kept quiet, whereas Oscar Levant in such circumstances was "at his wordiest and best." On one program the panel was asked to give a common household expression. Franklin P. Adams said: "Please pass the salt," while John Kieran came up with: "The front doorbell's ringing." When it came to Oscar Levant, he got the laughs with: "Are you going to stay in that bathroom all day?"

FPA could be quite devastating with his own impromptu wit. When someone was identified on the show as "his own worst enemy," Adams remarked: "Not while I am around."

Once the panel had to answer a question about a title which involved servants. "How Green Was My Valet," FPA replied.

Don't Say the Word

Oscar Levant had a phobia about death; he could not stand the mention of funerals, coffins, cemeteries, or anything to do with dying. He was approached in 1949 to appear on the quiz show *Who Said That?*. Levant agreed to try it. Fred Friendly, the producer, personally went to fetch the composer, and on the way to the show said to him:

"Oscar, you'd better be on your toes tonight, because we've dug up some cracks you made years ago to try and stump you."

"Where did you find them?"

"In the NBC morgue," said Friendly casually, referring to the extensive files that all news organizations have buried in their basements.

Levant turned white as he grabbed Friendly's arm:

"Never say that word to me or our association is at an end."

Pause for Thought

Broadway playwright Abe Burrows, appearing on a television panel once, was trying to guess the identity of a mystery guest.

"Is he living?" he asked for a clue.

"No," the quiz master replied, "he is dead."

"Let's see now," Burrows paused to think, "who do I know who is dead?"

A Brain You Can Trust

Dr. Jacob Bronowski, famous for his BBC series *The Ascent of Man*, appeared on *The Brains Trust*, a phone-in program on the BBC, on which experts answered the viewers' questions. On one program, the panel was grappling with a question sent in by a listener: "Does a chicken know it's a chicken?"

Sir Julian Huxley first tackled the question with a long but rather vague explanation. Then came a philosopher who similarly got himself bogged down in his own argument. Finally it was Jacob Bronowski's turn, and he said: "I don't know if a chicken knows that it is a chicken. I've never been a chicken. Who cares, anyway? What's the next question?"

(The English actor Anthony Hopkins was so impressed with Dr. Bronowski's straightforward handling of such questions that he became an ardent fan and corresponded with the scientist. As a result, Mrs. Bronowski chose the actor to introduce *The Ascent of Man* on American public television, after her husband's death.)

Assembly Line

Richard Dawson, when he was host of *Family Feud*, one of the all-time favorite game shows on American television, used to do his own warming-up of the audience before the show. Quick-witted, the British-born Dawson, who had once been married to Diana Dors, enjoyed the spontaneous give and take during impromptu interviews with the families before they started the competition. One time, he was talking to two sisters who worked making television sets for Panasonic.

"I spot weld them and hand them on," said one sister.

"And what do you do?" the quiz master turned to the other girl.

"I screw 'em," she laughed.

The audience was already laughing, but Dawson managed to look straight into the camera, and repeated the famous advertising slogan the electronics company was using at the time:

"Panasonic—just slightly ahead of our time."

It's a Challenge

The longest-running quiz program in Canada began in 1957 and is called *Front Page Challenge*. The title is descriptive: a panel of well-known journalists quiz a hidden challenger who is connected in some way with a front-page news story. After the panel succeeds or fails in identifying the celebrity within the given time, the second part of the program is given up to a face-to-face conversation with the challenger.

It is here that fireworks sometimes happened, especially in the days when the late Gordon Sinclair, a crochety no-nonsense Scot, was still a panelist. In 1969 he was almost fired after asking Canadian swimming star Elaine Tanner whether she swam during her period. The teenager calmly replied that she did, but more than three thousand irate viewers expressed outrage—a large number in the usually sedate Dominion.

The following season, feminist Kate Millett was lecturing the panel about the way some societies mutilated women, by binding their feet or pushing their breasts into bras. Gordon Sinclair listened for a while and then charged in:

"Well, they circumcise little boys before they even leave the hospital, and that's a form of male mutilation."

After the show, Sinclair learned that Kate Millett might know something since she had earned a Ph.D. "on the feminist subject." She complained that he had hardly given her a chance on the show.

"Okay," said Sinclair, "let's make up. If you show me your Ph.D., I'll show you my circumcision."

Nostalgia Isn't What It Used to Be

One of the hidden guests on *Front Page Challenge* in the winter of 1970 was Sally Rand, who had been famous in the thirties as a burlesque fan dancer. The show was being taped in Halifax, Nova Scotia, during a blizzard, and Miss Rand's plane was late coming in from Winnipeg, where she had been performing her fan dance routine—at the age of seventy.

"Finally, out of the storm came a lone taxi up the deserted street," reminisced Gary Lautens, a writer on the show and later columnist for the *Toronto Star*. It skidded to a stop, the door opened and out stepped a tiny old woman. She took one look at the drift of snow between the entrance and herself, hiked up her skirts and ploughed through."

The staff welcomed her effusively, and Don Brown, the producer, praised her spirit in the old "show must go on" tradition. Then noticing the ostrich plumes Miss Rand was clutching in her hands, he said:

"And you've even brought your fans. What a wonderful touch of nostalgia."

The little old lady brushed past Brown and Lautens, muttering:

"Fuck nostalgia. Where's the dressing room?"

Guess Who's Coming to Your Show?

Sam Goldwyn was invited to appear as a mystery guest on *What's My Line?*. The whole point of the popular game show was that the panelists had to guess the profession of the hidden guest by posing a series of leading questions. The movie mogul, who was in New York for the opening of his latest picture, failed to understand the niceties of the game; he thought of it as just another way of publicizing his new film. So when he saw one of the panelists, Dorothy Kilgallen, having dinner at "21," Goldwyn rushed up to her:

"Guess what, Dorothy? I'm going to be on your show Sunday night."

"Oh, Sam," the journalist reproached him, "you shouldn't have told me. Now I'll have to disqualify myself."

A couple of days later, Goldwyn ran into another panelist, the publisher Bennett Cerf.

"Did I do a dumb thing the other night," the producer confided in him. "I saw Dorothy Kilgallen at 21 and I told her I was going to be on her show this Sunday."

By the time of the show, most of the panelists had to disqualify themselves.

Goes With the Territory

Trying to guess the occupation of contestants often caused unintended hilarity on *What's My Line?*.

Bennett Cerf recalled the time columnist Dorothy Kilgallen broke up a sword swallower with the question: "Is yours an inside job?"

Stripper Gypsy Rose Lee was questioned: "Are you famous for your beautiful clothes?"

Ernie Kovacs asked a skunk grower: "Would I give your product to my girl for Christmas?"

And Cerf himself could not understand why he got such a huge laugh with his question: "Does your work cover a large territory?" until he found out that the contestant was in the diaper service business.

If Looks Could Kill

In order to promote his recently completed film, *Around the World in Eighty Days*, producer Mike Todd appeared, accompanied by his bride of two weeks, Elizabeth Taylor, as a mystery guest on *What's My Line?*. Listening to the studio applause given the flamboyant showman, panelist Bennett Cerf began his line of questioning which seemed very near the mark:

"Would our guest be, by any chance, the man who is the producer of what will be the box-office champion of this year and maybe all years?"

Todd smiled with as close an approximation of modesty as he could muster, and waited for the revelation.

"The man I have in mind is the producer of *The Ten Commandments*," Cerf went on and astray, referring to Cecil B. DeMille, who had just produced his remake of the biblical epic.

Mike Todd covered his face in frustration, while the audience roared. Since Cerf had not guessed correctly, the next panelist, singer Arlene Francis, quickly closed in:

"I'd like to remind our guest that one of those commandments is 'Thou Shalt Not Kill.'"

Unusual Program

Eddie Cantor, who had the reputation of always needing money, especially after the stock-market crash had wiped him out, once had an idea for a radio quiz show. He would select a name at random from a telephone directory, and if he found the person at home, Cantor would ask to borrow twenty dollars.

Guess What?

An elderly woman appeared as a contestant on a program which had Groucho Marx as guest. "Are you seventy?" he tried to guess her age. She shook her head. "Sixty? Fifty?" Groucho persisted. "You're getting very warm," the woman admitted coyly. "Not if you're fifty, I'm not," said Marx.

A Question of Choice

Groucho Marx was appearing on the popular radio program *We the People*, which specialized in "real people" and bits of Americana. One guest was a Mrs. Story from Bakersfield, California, the mother of twenty-two children, all from the same husband.

"Why do you have so many children?" Groucho began to question her. "That's a big responsibility and a big burden."

"Well," the big, pleasant woman explained, "because I love children, and I think that's our purpose here on earth, and I love my husband."

"I love my cigar, too," Groucho gave one of his celebrated replies, "but I take it out of my mouth once in a while."

In His Own Guise

When Groucho Marx began working on radio, after the Marx Brothers movies of the thirties had faded, he was without not only his brothers but also the landmark moustache. Bernie Smith, his producer, became worried that without the familiar image, studio audiences would not react to the comic.

"Groucho," he begged, "you have to wear your frock coat and the fake moustache."

"The hell I will," said the comic. "That character's dead. I'll never go into that again."

"Well," Smith pondered, "will you grow a real moustache?"

"That I will do," Groucho replied with Marxian logic.

Groucho Marx became a huge hit with a whole new generation of radio, and then television audiences. In January 1949 he was named the best quizmaster on the air. President Truman's surprise victory over Thomas Dewey took place just a few weeks earlier.

"It just goes to show," Groucho observed, "that a man with a moustache can get elected."

Maculate Concepts

Before he grew back his moustache, the producers of Groucho's quiz show *You Bet Your Life* came up with a sight gag for its feature, "What's the secret word?" At the beginning of each show, a wooden duck would appear revealing the word to the audience. The duck had thick, bushy eyebrows, wore glasses, sported a cigar—and a moustache.

With the success of *You Bet Your Life*, "What's the secret word?" became a catchphrase. According to Groucho, he was vacationing in Rome when somebody on the street knocked one of his large, expensive cigars out of his hand.

"Jesus Christ!" the comic exclaimed and tried to retrieve it from the gutter. But it turned out that the culprit had been a visiting priest from Cleveland who was now offering Marx a couple of fresh cigars:

"Groucho," he said, "you just said the secret word!"

In another celebrated encounter with Catholic clergy, Marx found himself in an elevator at New York's Plaza Hotel.

"Aren't you Groucho Marx?" asked the padre.

"Yeah."

"My mother is crazy about that quiz show you do," said the man of the cloth.

"I didn't know you fellows had mothers," Groucho rejoined. Then seeing the priest blush, he tried to extricate himself:

"I always thought it was Immaculate Conception."

Groucho used to tell about a priest he encountered in Montreal.

"Groucho," the cleric greeted him, "I want to thank you for all the joy you've put into the world."

"Thank you, Father," said Groucho, shaking the outstretched hand. "And I want to thank you for all the joy you've taken out of it."

Good Bet

One of the oddball contestants on *You Bet Your Life* was a writer who disguised himself as an Arab sheik. When Groucho Marx asked him what he was going to do with his share of the $10,000 prize, William Blatty replied:

"Finish my book."

He did. It was called *The Exorcist*.

Back to Oz

Mrs. August Jueneman, a farmer's wife from Rexford, Texas, had won $2500 on the original *Name That Tune*. The host, George de Witt, asked her if she would like to come back to try for $5000, and was shocked to hear her say:

"No, I have to be back tomorrow to sing in the church choir." Despite the producer's entreaties after the show that she should reconsider, Mrs. Jueneman returned to Kansas, and a substitute team had to win the next round for her.

Cold Cut

In the mid-fifties a housewife from Brooklyn, Rose Schmetterling, won jewelry and toys on *Be My Guest*, only to be told that the show had run out of actual prizes. She was offered instead a certificate to buy an equivalent amount in cold cuts. Finding that no market would honor the certificate, Mrs. Schmetterling went public with her opinion about the quiz show.

"Baloney," she said.

Winners and Losers

Arthur Godfrey used the same set for his television show as the quiz show *Winner Take All*. During one of his shows, he noticed that a washing machine, obviously an unclaimed prize, was left standing in one corner of the set. Godfrey asked a woman in the first row of the audience to give her full name. She did so.

"You're absolutely correct," Godfrey said, "and you win this washing machine."

When producers Mark Goodson and Bill Todman came in the next day for another round of *Winner Take All*, they found themselves having to pay for another washer.

Mousical Comedy

In the thirties radio stations or networks vied to come up with some ridiculous contests to keep audiences involved. Once the National Broadcasting Company organized a singing mouse contest which attracted entrants from Canada and Great Britain. The Times in London covered the event on May 3, 1937.

The broadcast yesterday evening, which was to decide the claims of England, Canada, and the United States to possess the sweetest-voiced mouse in the

English-speaking countries, found time for a flourish of trumpets, a song in honor of the contesting mice, and a good deal of chaff on the part of the announcers. Beside ourselves with pleasurable anticipation, we waited for the Canadian entrant to begin. But John, alias the "Toronto Tornado," was temperamental and refused all offers. Not a puff, not a peep out of him.

The English entry was a duet between Mickey, of Devonport, and Chrissie, a Welsh mouse, which goes to show that national talents are not confined to men. This would have been, in any case, a tour de force, since other countries were putting up only solo, or egotistical mice, mice unpracticed in the niceties of teamwork. Safe in the knowledge of their unique position, Chrissie and Mickey might have given some mediocre performance and let it go at that. But did they twitter half-heartedly, like mice who know their name is made in advance? They did not. They were British mice, they were artistic mice, they were mice of sensibility. So they piped away merrily in the most subtle harmonies, not a whit self-conscious. You just could not tell them apart, and that is saying a lot for duettists.

America had trouble too. Minnie, from Illinois, had been a glutton for exercise lately, probably deliberately, like film stars in the bad old days when it was a convenient way of being hors de combat at the beginning of a picture they did not wish to make. Anyhow, she merely ran around and around and refused to open her mouth. Mikey, from the same state, made up for her. Here was mouse music at its gayest, with not a trace of those intimations of mortality one detects in the lower notes of crooners. All Mikey's notes were high ones, delivered with such virtuosity and vigour that one sees in him the coming mouse Caruso.

This is not, we may be sure, the last we shall hear of the singing mice. Mouse opera has been suggested already, and there is no reason why recitals by mice should not be popular. There will not be, one hopes, mouse crooners.

Prattling Polly

One bad idea will readily spawn another, and NBC soon followed its singing mice contest with one featuring talking parrots. In his book I Live on Air, *A. A. Schecter, who organized both contests, described the field.*

Not long after our contest was announced we were informed that the following were among the world's blessings that we could put within our reach by just saying the word:

1. A parrot that could recite the Lord's Prayer.
2. About a dozen parrots that could imitate Charlie McCarthy.
3. Four or five that could imitate a train announcer.

4. A parrot that could imitate a baseball umpire. (His specialty was saying, "Strike tuh!")
5. One that could say, "Papa's home! Papa's home!"
6. One that could bark, mew, moan, crow, moo, and gurgle.
7. Several that could say, "Skoal!," Bottoms up!," "Down the hatch!," etc.
8. A great many—I don't recall the number—that could whistle.
9. One that could hiccup.
10. One that could belch.
11. At least a score that could give the Bronx cheer.
12. One that could say, "Fish!—nice fresh fish!"
13. Five that could sing *The Sidewalks of New York.*
14. Three that could sing *Sweet Adeline.*
15. One that could sing *Frankie and Johnnie.*
16. One that could yell, "Down in front!"
17. One that knew most of the strange lingo of swing-music fans.
18. One that could say, "For further details, read your daily newspaper."
19. Two that said, "How do you get that way?"
20. Three that said, "Time please, Central."
21. One that said, "Your slip's showing, Mabel."
22. One that cried, "Nice fresh peanuts!"
23. One that talked Yiddish.
24. One that exclaimed, "Hooray for the Irish!"
25. One that recited, "We shot our youngster dead 'cause he wet the bed."
26. One that said, "There's a fly in my soup."
27. About three hundred that said, "Polly wants a cracker!"
28. One that could croak a French pun, to wit: "Polly vous Français?"
29. A half-dozen or more that could say, "Go jump in the lake!"
30. One that yelled, "God save the King!"
31. One that said, "Haul in the halyards!"

. . . Mr. Cuppy, Chairman of the Committee of Judges, made this momentous pronouncement with the comment: "I'm glad I'm on the judging end of this thing. I certainly wouldn't want any parrots judging *my* diction."

The Ultimate

Some countries have evolved game shows into major events, rather than the strictly controlled thirty-minute programs on American or British television. Several Western European countries collaborated for years on staging arduous, pseudo-Olympic events, in which teams from various nationalities competed in the *Games Without Boundaries.*

The Japanese have created *The Ultra Quiz*, a brutal and humiliating combination of boot camp and nightmarish exams, which reduce 5000 contestants, after three weeks of elimination rounds, to just two finalists. As shown on an ITV program about television, one particular game took place on airplanes and cruise ships across the Pacific, until the last two survivors fought it out on top of the PanAm skyscraper in New York City.

Michiyo Saito, creator of *The Ultra Quiz*, which he has called a three-week documentary, thinks that the enormous popularity of the program feeds off the Japanese predilection for extreme situations. The time would come, he predicted, when the monstrous and insatiable audience, like in ancient Rome, will demand that someone be killed on a game show.

Take My Co-Host, Please

In its attempt to reach farther and farther out, The Garry Moore Show *decided to go for a novel twist on a giveaway contest—as producer Herb Sanford chronicled the event.*

In the spring of 1954 we announced that Durward Kirby would be given to the writer of the best letter stating "Why I would like to have Durward Kirby for my very own." Ownership was of course only for a weekend. We needed him back in time for Monday's show. The plan was to crate Durward and ship him to the winner, express charges prepaid.

Everything was set in motion. Formal request for the crate was issued to Set Construction. John Neukum, our assistant producer, began negotiations with Railway Express. The Set Construction people appeared bemused; they thought we had finally crossed the line. They decided to humor us and asked about specifications. Dimensions would need to be sufficient to hold Durward with some degree of comfort. The crate would have to be sturdy enough to withstand the journey in a baggage car. And there must be air holes. Also it must look smart, because we planned to display it on stage.

Railway Express told us there were no regular rates for this sort of thing. After all, there was no precedent. But they were indulgent; they agreed to see what could be worked out. They were not sure we would actually bring it off, but, as John Neukum remarks, "they thought it was so bizarre they joined in."

The network publicity people gave us a blank stare. This was surprising, since we thought they would go for anything that would make news. They said we couldn't get away with shipping a live person. But the story went out to the papers and the papers printed it.

Bales of letters began arriving—so many that it was difficult to manage safe passage through the space adjoining our offices. Some of the letters were interesting and well written. It was evident there were a lot of housewives who would like to have Durward Kirby for their very own.

After the announced closing date, several weeks were required to pick out a hundred or so to be considered seriously. Garry, along with others of us, found time to read all of these. Before announcing the winner, we conducted our own private investigation to make sure all was well. Not a private-detective type of investigation. We simply asked the local station to give us a rundown on the chosen winner. It would have been embarrassing to find we had picked a shady character who knew how to write a good letter.

We were lucky. The winner was Elly (Mrs. Robert) Morse of Chardon, Ohio, near Cleveland. Mr. Morse was a construction engineer, owning his own company in Cleveland. Garry telephoned the news to Mrs. Morse, so that she could plan on her house guest and tune in the show to hear herself announced as a winner. Mrs. Morse was pledged to secrecy (except for Mr. Morse) so that there would be no leak ahead of the announcement.

We gave Durward a send-off that Friday on the show. Garry, Durward, and a representative of Railway Express stood alongside the crate, which had one panel open. Garry announced the winner, everyone wished Durward good traveling, Durward stepped inside, the crate was sealed, and Durward was on the way.

The crate was given preferred position in the baggage car, and Durward was allowed periodic visits to the club car and a Pullman berth for overnight sleeping. The next morning, the crate was put off at Cleveland. It was abandoned at that point; it would have been asking too much of Mr. and Mrs. Morse to carry it home. Durward was delivered at the door of the Morse home by bonded messenger.

The phone in the ranch-style house was ringing as Durward arrived. It never stopped. Some of the calls were from complete strangers, who asked if they could borrow Durward for ten or fifteen minutes. Mrs. Morse transferred the action to their place in the country, where the phone was unlisted.

That evening, about twenty friends came over for a buffet supper. On Sunday, Durward's wife, Pax, came over from Indianapolis, where she was visiting, and joined Durward and Elly and Bob for dinner.

"The weekend turned out great," says Durward, "but it could have been a funeral for me. You see, *Mr.* Robert Morse had never heard of Durward Kirby in his entire life! So you can imagine my consternation at being confronted by this gentleman in his own living room when he returned home from work, with these words: 'Bob, I'd like for you to meet Durward. I won him for the weekend on *The Garry Moore Show.*' I didn't know for the moment if he was going to shake my hand or wring my neck!"

Mrs. Morse also remembers Garry calling and telling her that she had won Durward. "Garry told me not to tell anyone until it was announced on television, but that I could tell my husband. Bob, of course, didn't even know I had entered the contest, as I had no idea that I would even be considered. When I called my husband at the office to tell him we would be having a weekend guest, he said, 'Who in the hell is Durward Kirby?' I explained it to him, and after a few choice remarks he decided we'd better do a little investigating of our own, as he wasn't about to entertain any of those 'damn show people.'"

Ethical Dilemma

Mike Wallace was on a quiz-show panel and asked a contestant: "What would you do if you found a million dollars in cash lying on the sidewalk?" The woman pondered this ethical dilemma for a moment and then replied thoughtfully: "If the million in cash had been lost by a poor family, I'd return it!"

CHAPTER 12

Blow by Blow

The Babe

During the summer of 1923, when the New York Yankees went to Pittsburgh for an exhibition game, a reporter brought Babe Ruth to KDKA studio for a radio interview. The great baseball slugger got such mike fright that he could not read the short speech that had been prepared for him. Harold Arlin read the script on the air instead and the station received several letters remarking on the broadcast quality of the Babe's voice.

Hog Wild

That same summer the boxing heavyweight championship fight between Jack Dempsey and Luis Firpo took place. As the only way to broadcast a fight before the days of remote or on-the-spot reporting, announcer Harold Arlin took the blow-by-blow description off the wire service and then re-created it for his radio audience.

At one point, when Firpo gave a blow to Dempsey, the wire went dead, and

Arlin instinctively switched to a report on farm commodity prices. Next day he ran into a friend who said:

"Well, Firpo knocked Dempsey into the ropes, and hogs went up two cents a pound."

Feedback

Long before they had Howard Cosell to kick around, American sports fans were not bashful about venting their feelings about the people behind the microphone. Graham McNamee received fifty thousand pieces of mail after giving the blow-by-blow at the 1925 World Series; up from seventeen hundred the year before. There were at most five million sets in use throughout the country. McNamee published some of the missives that tickled him most.

A Pittsburgh fan wrote me: "We would appreciate it if you would not constantly remind us that the score was Washington 4; Pittsburgh o."

Another appealed to the office: "For the love of mike" (an unconcious witticism), "take McNamee out. My batteries are valuable and I can get jokes from the newspapers."

The Washington fans took the series quite as much to heart. Wrote one:

"You utterly and miserably failed in your rôle. You announced, yes, but spent more time telling about your personal discomforts (as if we cared a damn) than you did about who was at bat and what he was doing.

"You failed in 1924, but, My God! you get worse as you get older! Stick to music and weather reports, but let the Chicago man handle the large athletic events for the good of us all."

An associate judge in the capital seemed to feel the same way, though he expressed his feelings in a little different fashion.

"It would have been excellent," he wrote, "had you been mindful of your audience and its divergent sympathies, and divested yourself of your very evident prejudice in favor of Pittsburgh.

"No doubt any complimentary letter will be exhibited to your employers. Will you be man enough to show them this too?"

Well, Judge, there it is.

Carried Away

Sam Taub, a veteran sports announcer for NBC in the thirties, became completely involved in some of the boxing fights he covered. At one point he was giving a blow-by-blow account when the boy he favored went down to kiss the canvas for the third time. The announcer gave vent to his disappointment in somewhat unprofessional terms. Collecting himself, Taub covered up by telling his radio audience:

"An uncouth person just passed behind me. I hope none of his remarks got into the microphone."

Immortal Words

Max Schmeling lost the heavyweight boxing title to Jack Sharkey in 1932. It happened on a decision rather than a clear knockout. Schmeling's manager, Joe (Yuseel) Jacobs, leapt into the ring and started berating the referee, when he noticed a radio announcer stalking his client to see if he could get the defeated champ to say a few words.

Jacobs saw his chance to make his protest to a larger audience; he made a beeline for the microphone and created an instant catchphrase: "Ladies and gentlemen, this is Joe Jacobs speaking, Max Schmeling's manager. We wuz robbed here tonight!"

Punchline

Rocky Graziano, world champion boxer, appeared as guest on *What's My Line?*.
    "Are you a pugilist?" Hal Block asked him.
    "Nah," Graziano replied, "I'm just a prizefighter."

Shakespeare Scores a K.O.

In his book I Live on Air, *A. A. Schechter, director of special events for NBC in the late thirties, describes an unusual radio broadcast of the Bard.*

Tony (Two-ton) Galento was more than a pretty good heavyweight boxer. He was an excellent actor and his talents as a Thespian were more instrumental in making him a fine drawing-card than his pugilistic ability. Some fighters resent being told they are acting. Tony, on the other hand, was always proud of his talents as an itinerant player. In many a city they had applauded his histrionics and he liked it.

Knowing of Tony's pride in his accomplishments as a dramatic artist we felt we were on safe ground in inviting him to play Romeo to Nancy Carroll's Juliet in the balcony scene from Shakespeare's romantic tragedy. Tony accepted and the coast-to-coast broadcast that resulted was cause for rejoicing among drama lovers.

Tony had been built up as a boxing attraction by means of goofy stunts, and there was never any doubt that he would accept. The Louis-Galento fight was only four days off; and, after all publicity is publicity.

Tony, however, gave me a few bad moments. He failed to show up for the rehearsal we had planned, and five minutes before the show was to go on the air he had not yet arrived. The page boys I posted at all entrances to the RCA Building, where our New York studios are located, reported they could find no trace of the boxer-actor. Their job was to rush Tony to the right studio the second they spotted him.

Four minutes before Milton Cross was to make the opening announcement Tony walked calmly into the building with his wife. The page boys grabbed him, rushed him to an elevator, and fairly pushed him into the studio where I was waiting none too patiently.

"What's the rush, Abe?" Tony asked. "I got three minutes, ain't I?"

In the studio, however, Tony lost a bit of his poise after he glanced at his script. He turned to sports announcer Bill Stern, who was to interview him later, and declared:

"Why didn't you tell me? I'd have gone in trainin'."

Galento's confidence returned, however, when a minute before the program got under way Stern asked: "Think you can handle Shakespeare, Tony? A lot of guys find his style baffling, you know."

"Shakespeare!" exclaimed Two-ton scornfully. "I'll moider dat bum!"

And moider him Tony did while a gathering of sports writers, drama critics, magazine editors and fellow Thespians looked on happily; for the boxer was raising bad acting to new heights. Jaded Broadwayites in the studio who on many an occasion had sat grimly through the pretty good kind of bad Shakespearian acting recognized Tony's efforts as bad acting at its very best and applauded enthusiastically when it was over.

From the standpoint of the radio audience it was desirable too that Tony be superlatively bad; this meant more fun for our listeners. If Tony had been only mildly lousy Vox Pop would have written in to say, "Why did you let that Galento play Romeo? There are lots of better actors. Frankly, I think he's a ham." There was no danger that anyone would so quickly graduate Tony to hamdom, a state enjoyed by many experienced performers.

Tony's lack of rehearsal helped his performance tremendously. When Nancy as Juliet declared: "If they do see thee they will murder thee," Tony replied: "I have night's clock to hide me from their eyes." He had, of course, garbled the word "cloak."

He also stumbled over several lines. "But I'll sock anyone," Tony said later, "who says dat proves I'm a stumble-bum actor. I just ain't fully trained. When you're booked to meet a guy like this Shakespeare you gotta do roadwork or them big woids get you winded."

Perhaps the hardest job of all was Bill Stern's. Bill's role was to interview Galento on his ring plans after the Shakesperian act was over. As the fifteen-minute broadcast neared its conclusion, Romeo and Juliet were still at it. Alas, Shakespeare's puzzling style was slowing up Tony, and the play-acting was taking more time than we thought it would.

There was nothing for Bill to do but to interrupt. And interrupt he did. "What," said Bill, cutting in as Two-ton addressed a poetic passage to fair Juliet, "have you got to say, Tony, that would interest the fight fans?" It was then that Tony made the historic training pronouncement that went echoing down the corridors of fistiana.

"I'm cuttin' out de beer," said Tony, who all through his ring career had trained on lager. "It sours me stomach. I'm goin' to malted milks."

Knock Wood

Floyd Patterson was being interviewed on *The Tonight Show* by its original host, Steve Allen, who wanted to know why the champion boxer had named his daughter Seneca.

"Well, my wife and I were trying to decide on a name for the baby," Patterson replied, "and one day we happened to be walking down this street in Brooklyn, and we looked up at the street sign and it said Seneca."

"You're lucky," Allen rejoined, "that you weren't walking on Flatbush Avenue."

Knockout

At another time Steve Allen was conducting an interview from the studio with Sugar Ray Robinson on location at a training camp in upstate New York. There was equipment trouble, and while it was being fixed, the producer asked Allen to say something into the microphone.

"One, two, three, four . . ." the comic started counting idly, until he reached ten. Then realizing that the boxer could probably hear him, he added: "Nothing personal Sugar."

Pearl Before Swine

It is common for former champions to become sports commentators or to lend their expertise to the announcer. Occasionally a celebrity will be asked to do the commentary, and then the results can be unpredictable. Don Dunphy, the veteran sports announcer, recalls the 1973 heavyweight boxing title between defending champion Joe Frazier and George Foreman in Jamaica. Super-promoter Don King asked Pearl Bailey to be the color commentator from the ringside. As soon as the fight began, the great singer forgot about the need for a commentator to remain impartial; in fact, she forgot about her whole assign-ment. As she cheered on Foreman, just like any fan, Bailey managed to block the view of all the other broadcasters and sportswriters. She ignored all of them yelling at her to sit down.

"Finally came the coup de *swat*," Don Dunphy observed in his book of memoirs. "As Foreman floored Frazier, Pearl screamed with joy and whacked me on the back with what she must have thought was a playful slap. I was knocked forward, and my ringside papers went flying. I still wince when I think of that jolt. I wondered who was defending the title, Frazier or I."

His Royal Heinous

The night before the heavyweight championship fight between Muhammad Ali and Leon Spinks in 1978, Johnny Carson remarked that people could see the boxers for free on television. "Well, not exactly free," Carson added. "You have to listen to Howard Cosell."

No television personality has endured the kinds of assaults and insults that have been Howard Cosell's lot since he gave up law and began broadcasting a Little League panel show in the late 1940s. His Brooklyn twang, polysyllabic vocabu-lary and aggressive questioning have made Cosell the most controversial—and watched—figure among American sports commentators. One critic said that "ten minutes of Howard Cosell is like spending 24 hours with Idi Amin." Cosell has been called everything from "old flannel mouth," "Erich von Stroheim," "Mr. Jaws," to "the Most Powerful Irritant Since Itching Powder."

Fans love to hate Cosell. During ABC's telecast of *Monday Night Football* banners and posters would be waved at the camera, like red flags at a bull, reading: "Will Rogers Never Met Howard Cosell," "Buffalo Without O. J. Is Like Cosell without a Toupee," or "Cosell is Un-Bear-Able." Once an elaborate three-dimensional banner, flapping in the wind, was a visual representation of "Motor Mouth." In San Francisco, ABC had to remove a banner which said: "Fuck you, Howard!"

Cosell's longtime boss, Roone Arledge, has a favorite story about a bar "Down South" where all the regulars gather on Monday night, having pooled some of their money to buy an old TV set and some buckshot. "They then draw lots and the first time Howard's picture comes on the screen, the winners get to blast the TV set to smithereens." Variations on this ritual took place weekly at bars in Colorado, Indiana, Michigan, and even Brooklyn.

At the 1977 World Series, an angry fan aimed a pair of pliers at Cosell; in 1975 a bunch of Minnesota Vikings made merry by dousing him with water, sweeping away his toupee. After the fight between Scot LeDoux and Johnny Boudreaux in the United States Boxing Tournament of Champions in 1977, somebody aimed an ice-filled cup at Cosell's head and almost knocked off his famous hairpiece.

Bad Press

The press hated Howard Cosell without the love. Once, when Cosell was coming down with the flu on the air and had to leave halfway through a game between the New York Giants and the Philadelphia Eagles, sports journalists immediately concluded that he must have been drunk. Informed that Cosell was suffering from an inner-ear infection, Dick Young of the *Daily News* retorted:

"Yeah, he poured three martinis in his ear."

Chatting With Howard

Most American athletes are used to Howard Cosell's provocative interviewing style. But he was dispatched in 1966 to England to talk to Brian London, before his fight with Muhammed Ali. Cosell and producer Joe Aceti finally tracked down the Englishman at a seedy training camp in Blackpool. Sitting next to London on a bench, Cosell put his arm around the boxer and opened in his usual manner:

"Brian, they're saying that you're a patsy, a dirty fighter, that you have no class, that you're just in there for the ride and fast payday, and that you have no chance against Ali. Now, what do you say to all that?"

London looked at his interviewer with a growing sense that he was being insulted.

"Go fuck yourself!" he replied simply.

"No, no," Aceti intervened. "You see," he said to London, "those really aren't Howard's descriptions—they were opinions expressed by others."

"Oh, I see," said London and offered to take the question again. When Howard Cosell repeated the statement word for word, the pugilist replied:

"Whoever said that can go fuck themselves."

Hair

Charles Kuralt, the lovable, balding Buddha of CBS News, was asked what he had found most impressive during his long travels for his whimsical series *On the Road*. Kuralt said that, watching the local news in a different town practically every night, he was struck by the hair on the local anchormen.

Hairpieces are essential props for television personalities, several of whom have made no secret about their lack of hair; Willard Scott, the irrepressible weather-man on NBC's *Today Show*, puts on a toupee some days and goes naked on others. Howard Cosell's hairpiece had featured in many a mêlée in the sports commentator's tumultuous career. Sitting next to him one day, comic Pat Henry said:

"I'm getting stoned just sniffing the glue on his hairpiece."

Bill Stern, the most famous sportscaster in the early years of radio, became the first commentator of a televised baseball game back in 1939. NBC management insisted that the veteran announcer must wear a hairpiece for this momentous occasion. But Stern, not used to wearing a toupee, which he hated, left it behind in his office. Now he had to turn back to retrieve it. By the time he turned his car again toward Baker Field, where Columbia was about to confront Princeton, Stern was speeding until a motorcycle cop pulled him to the side. He began to lecture the errant motorist, who was desperate that he would miss the beginning of the game that he was supposed to introduce.

"Officer," he begged, "you won't believe this story. I am hurrying so I can broadcast the first baseball game ever shown on television."

The cop had never heard of television. Stern went on to explain how he had to go back and get his toupee from the RCA building. The policeman looked dubiously at the inert piece of hair resting in a box next to the announcer. The story was becoming curiouser, and the peace officer began to think that perhaps the man needed to see a doctor. Finally, Stern identified himself:

"Officer, please, take a closer look at me. I'm Bill Stern, the sportcaster. Please give me a break."

Stern made it to the stadium just in time for the historic broadcast.

Sending a Message

Sometime in the late thirties, Bill Stern was giving the play-by-play for a game between Notre Dame and USC in Los Angeles. One of his jobs was to give the scores from other matches as they came over the wire from Western Union. Stern noticed that the wire service gave the wrong score on the game he was covering, so he persisted in giving what he believed to be the correct one. Western Union was listening, because it started sending messages to warn Stern that he was broadcasting the wrong score between Notre Dame and South Cal. After the announcer ignored several of these, finally a message came over the tape:

"This is Western Union. Can't you read?"

Two Kinds of Facilities

In the days before exclusive broadcast rights, both NBC and CBS decided to cover the 1939 football match between Alabama and Vanderbilt in Nashville.

Vanderbilt Field had no broadcast facilities, so the college administration agreed to set up two booths at the top tier of the stadium. They were narrow, temporary structures, and some pranksters painted "MEN" on the NBC booth and "WOMEN" on CBS's studio.

Throughout the game, men and women from all over the stadium were trying to fight their way inside these booths, only to find them already occupied by announcers.

Kids Say the Darnest Things

Ted Husing, one of the famous sports commentators on early radio, made the mistake of giving the microphone once to his five-year-old daughter at half-time in a football game between Manhattan College and St. Bonaventure.

"Good afternoon, everyone, everywhere," chirped the child in imitation of her father. "This is Peggemae Husing. Daddy calls me Pedge. Manhattan has the best team and the prettiest band uniforms. I certainly hope they win for my Daddy's sake, because I know he and Mummy had a five dollar bet on this game."

Ted Husing wrote in his autobiography that his stunt "was never done before, and I'm reasonably sure it won't be attempted again."

Tea Party

Ted Husing was interviewing Sir Thomas Lipton, the English tea producer, who also had a reputation for being something of a wit. The sportscaster first asked the veteran yachtsman about his chances to win that year's America's Cup race.

"Is there any of the four boats that you fear?"

"No, young man," Sir Thomas replied. "I'm not afraid of any of them."

"The experts believe," Husing persisted, "the *Yankee*—the boat from Boston—to be the fastest."

"Boston, Boston," the baronet mused, "O, yes, Boston. Wasn't that the city where they had a tea party a long time ago?"

"That's correct," the American announcer replied, "but if you'll pardon me, Sir Thomas, it wasn't your tea."

A Day at the Races

John Snagge became world-famous for his BBC commentaries on the Oxford and Cambridge boat race for fifty years. In 1949 in a desperately close race he made his best-known gaffe: ". . . It's a very close race—I can't see who is in the lead—it's either Oxford or Cambridge!"

A woman was watching preparations for the annual Oxford and Cambridge boat race near the BBC commentators' booth. "I wonder if you can explain something to me," she asked Tom Boswell, one of the announcers. "I come to see the boat race every year. How does it happen that the same two crews are always in the final?"

Brian Johnston also recalls an "unknown" commentator who called the race one year between two Oxford colleges at Henley Royal Regatta: "It's a very close race. Lady Margaret and Jesus are rowing neck and neck. Perhaps Lady Margaret is just ahead . . . but no, Jesus is now definitely making water on Lady Margaret!"

Wrong Call

Harold Abrahams, one of the two runner heroes of the movie *Chariots of Fire,* where he is shown winning the gold medal for the 100 meters in the 1924 Olympics, went on to become a well-known sports commentator in Britain.

In broadcasting the finish of the 1955 marathon race from Windsor to Chiswick, he called a runner named Iden as the winner, having failed to notice the lead runner already in the stadium. When the mistake was pointed out, Abrahams apologized for the case of mistaken IDENtity.

Don't Adjust Your Set

Sports commentary on the BBC is much more subdued than the frantic mixture of instant replays, statistics, constant analysis, and commercial breaks that characterizes American television. Dan Maskell, the veteran British commentator, recalls in his memoirs how one year at Wimbledon the NBC team switched to the BBC coverage. Suddenly, all that could be heard during the forty-five seconds between the playing of each point were the natural sounds in the stadium; Maskell made no comments "simply because there was nothing to say that would enhance the broadcast for the viewers." Bud Collins, the NBC commentator who is rarely inhibited by such considerations, kept breaking in to reassure American viewers:

"Don't worry, folks, we really are tuned to the BBC commentary—they just don't happen to be saying anything at the moment!"

Big Bad John

The Davis Cup final in 1978 between Britain and America was broadcast from Palm Springs on BBC radio. The team of commentators included veteran Dan Maskell, a former tennis champion and coach, who for many decades was the voice of Wimbledon. He and his colleagues, Gerry Williams and Max Robertson, were perched in a cramped structure on top of the clubhouse, and sometimes their commentary could be heard by the players.

Predictably, young John McEnroe, hearing the voluble Dan Maskell describe the scene, looked up and in his inimitable manner screamed, "Shut up!" at the venerable gentleman. The British, who would not be exposed to the full McEnroe treatment until his Wimbledon victories in the eighties, were understandably scandalized. Gerry Williams was still shaking his head years later, recalling the incident on the television program *This Is Your Life* when it honored Dan Maskell.

"Imagine," he marveled, "anyone expecting Dan to shut up!"

Amateur Time

In the early days of commercial television in Britain, the man in charge of sports for ITV came to Lew Grade one day and informed him breathlessly that he could get film of the Olympic Games.

"I am not interested," said the mogul.

"But these are the greatest amateurs in the world!" said the man, deeply shocked and disappointed.

"Amateurs!" Grade screamed at him. "I'm only interested in professionals!"

Bowl Movement

NBC and ABC were locked in a fierce battle for the rights to televise the 1968 Winter Olympics in Grenoble. NBC, by far the richer applicant at the time, put on a dazzling multimedia display which placed great emphasis on the network's experience in producing such big American sports events as the Rose Bowl, the Orange Bowl, and, of course, the Super Bowl. Despite a much more modest presentation, ABC won the rights. Roone Arledge, head of ABC Sports, was puzzled, until he received congratulations from the chairman of the Grenoble Organizing Committee.

"And please, could you help me?" the Frenchman added. "I want to know why NBC kept talking of the Bowel Games? It was in very questionable taste."

You Say Hello

Roone Arledge had to discuss some complicated technical requirements with the Austrian Organizing Committee in preparation to televise the 1976 Winter Olympics at Innsbruck. He suggested that an interpreter be present, but the committee chairman declared that his English was excellent and that he understood everything perfectly. The head of ABC Sports was a little dismayed when after four hours of detailed technical presentation, the chairman stood up and said:

"I see, by looking at my watch, that it is time to say hello."

Wide World of Sports

Geoff Mason arrived in Innsbruck to prepare ABC television's coverage of the 1976 Winter Olympics. Concerned about oversleeping because of jetlag, the producer instructed the front desk of his hotel to wake him at 6:30 and, just in case, to give another call at 7 A.M.

"Fine, sir," said the voice at the other end, "which call do you want first?"

Jim McKay, the ABC sports announcer, was part of a preliminary press tour of the new Olympics facilities in Montreal in the summer of 1976. Taken to see the Velodrome, McKay was the only one in the group who claimed to know the derivation of the word.

"Velo," he explained, was related to velocity, meaning speed; "drome" came from dromedary. "So a velodrome," McKay concluded, "is a place for fast camels."

The Real McKay

The real name of Jim McKay, one of the best-known sports commentators in America, was Jim McManus. He was renamed in 1950 when he went from Baltimore to work for CBS. In the old days of radio, a station or a network owned certain names which ensured continuity if a particular announcer quit. Jim McKay helped to destroy that system when he left CBS for ABC and took his new name with him.

The Grand Compromise

In its early days, ABC's *Wide World of Sports* was in Acapulco to report on the International Water Ski championships when it discovered a local sideshow—the spectacular divers from ninety-foot cliffs into the bay. Producer Roone Arledge and director Dick Kirchner wanted to know how much it would cost to get these amateurs on tape. Their local guide, Raul Garcia, told the gringos it would be a hundred thousand dollars. It was a huge sum for a bit of local color, and there seemed no point in bargaining. Seeing the deal slip through his fingers, Garcia asked for a moment to consult with his clients.

"It's all right," he came back with a triumphant smile. "They compromised."

"What did they compromise for?" asked Arledge.

"They've agreed to ten dollars a dive," said the negotiator, "including each rehearsal dive."

That is how the Acapulco cliff-divers became regulars on *Wide World of Sports*.

Cliffhanger

Roone Arledge revolutionized the telecasting of sports in America by letting the viewer at home experience the event through the camera and commentators. One of his early insights, that sports must have appeal beyond hard-core fanatics, came while watching the annual football game between the Army and Notre Dame in the mid-1950s. Only thirty seconds remained in a very close game. As one team lined up for what would be a decisive field goal, Arledge's wife, who was generally indifferent to football, asked for the binoculars. Surprised, he asked her what she was looking at.

"The Notre Dame band has gold tassels!" she exclaimed.

Timeliness Is Next to Godliness

In the early days of telecasting basketball games on Sundays, the National Basketball Association was so careful not to offend any religious groups that it did everything to finish its games on time. "You could tell those guys," said an NBC executive, "that they had to be off the air by 3:58 P.M. because the *Catholic Hour* or something was coming on—and they would be off at 3:58 P.M. They usually did it by refusing to call time outs if they looked like they might run over."

On the other hand, Roone Arledge believes that television has ruined golf on the public links. "Everyone is now taking three minutes to line up every shot," he told journalist William O. Johnson, Jr., "because they saw Jack Nicklaus do it on the tube. *That* is what I call an impact on sports."

Masking

Television has changed forever the nature of sports, and not just because of the money. As Larry Colton, the former pitcher with the Phillies who went into television, once put it: "It's not whether you win or lose, but how good you look in your uniform." The vanity of being seen by millions of people has inflated the egos of many participants. Masking is a technical term in the theatre for when one actor gets in front of another, blocking the audience's view. This was never a problem in fast-moving sports. But when quarterback Sonny Jurgensen, playing for the Washington Redskins, once ran up to referee Jim Tunney to protest a call, the umpire told him:

"Sonny, get out of my way; we're on national TV and you're standing in front of me."

CHAPTER 13

Making a Mint

The Selling Machine

The awesome power of selling products on television became manifest with the phenomenon of the quiz shows in the mid-1950s. Revlon was an upstart among cosmetics companies when it decided to underwrite *The $64,000 Question*. The show had been offered to Helene Rubinstein, who did not own a television set and believed that "only poor people watch those awful machines."

Charles Revson was also opposed to advertising his colorful wares on black-and-white television, but felt forced into it by a competitor who was stealing market share by sponsoring *This Is Your Life*. Still, he would commit Revlon to only thirteen weeks. Executives of both Revlon and CBS were astonished when *The $64,000 Question* became the number-one show on television within four weeks. It built to a point when an astonishing 82 percent of the sets in America were tuned into the show every Tuesday night. The impact on Revlon's sales was no less monumental. The live commercials, supposed to be sixty seconds long, often went on for 2–3 minutes, increasing the sales on some Revlon products between 300 and 500 percent. It was not uncommon for a shade of lipstick to sell out the morning after the show. As one Revlon executive delicately phrased it to writer Andrew Tobias: "We could have sold urine in a bag."

Bought and Paid For

During the 1960 congressional hearings into the fixing of the game shows, Charles Revson was asked whether he thought of returning any of the profits his company, Revlon, made from the immense success of *The $64,000 Question* or *The $64,000 Challenge*. Revson did not see why he should.

"We paid for the show," he told the committee. "We paid for the time. We paid for the contestants, sir. So, therefore, in turn, we made a profit on it."

Patronage

The patron saint of television—and God only knows, it needs one—is St. Francis de Sales, just as the other St. Francis, of Assisi fame, is considered the protector of journalists. Nobody quite knows how the Vatican gave this assignment to St. Francis de Sales, but one pragmatic American executive guessed that it came about because "there is no television without sales."

Paying the Piper

Comedian George Gobel, in the early days of the medium, would ease his way into a commercial break: "You know, commercials are the life blood of television, and this anemic industry needs all the blood it can get—so please watch this."

During his series of suspense dramas, *Alfred Hitchcock Presents*, the great director and ham performer tried to find ingenious and natural ways of leading his audiences into a commercial.

"When I was a lad," Hitchcock narrated, "I had an uncle who often stood me to dinner. He always accompanied these dinners with interminable stories about himself. But I listened carefully because he was paying the check. I don't know what reminded me of my dear old uncle, but we are about to have one of our commercial breaks."

Hucksters

One of the American news commentators in the 1930s, Boake Carter, perfected the art of combining public affairs with commercial copy. He became a huckster for everything from Post Toasties to Nash automobiles. At one point, his broadcasts were sponsored by Philco, the makers of radio sets. Typically, Boake would give listeners the latest news from Europe in the following manner:

"The Communists became enraged. Chautemps collapsed. So the picture of

Europe bubbles again—and becomes a point of interest again for those who like to tune long distances on their Philco 116 Double Xs. One does not have to double oneself in knots as of old to see if you're tuned just right. That's the pleasure of a Philco Double X."

Another well-known radio commentator in the thirties and forties, Gabriel Heatter, became even more adept at introducing commercials, which, according to one critic, he slipped "into his news discussion with the deftness of a Sicilian footpad easing in a knife." Another attacked Heatter in yet harsher terms: "His equal anxiety over Dachau and dandruff, Japan and gingivitis made his news commentaries ludicrous."

Gabe Heatter worked on the preparation of his broadcasts with his brother Max. One reported conversation went like this:

"What is the product for tonight?"

"Kreml shampoo, Gabe," replied his brother.

"Impossible to follow the Burma jungle item with a shampoo," the broadcaster declared. "It will have to come after the Polish question."

"Definitely," Max agreed.

Broadcast Forecast

WJZ, one of the pioneer radio stations operated by Westinghouse in Newark, New Jersey, broadcast in the early twenties concerts by Vincent Lopez's orchestra. After a while, Lopez grew tired of having to cross the river with his players from Manhattan; he suggested that WJZ should come to him and transmit live his nightly performances at the Pennsylvania Grill.

No sooner were the concerts announced on the air than the restaurant was inundated with fans who wanted to be part of the ambience. Owner E. M. Statler, of hotel fame, watched with incredulity his reservation book fill up night after night, and was finally moved to prophesy:

"Vincent, I couldn't build business up like this in a thousand years of hard work. You did it in an hour. I think radio has some real possibilities."

Yes Cigar

Radio as an advertising medium almost became a victim of its own success. Credo Fitch Harris, manager of WHAS in Louisville, recalled one of the early "time" salesmen visiting his station in 1925 to offer sponsorship of programs— $400 an hour—for his client, a cigar manufacturer.

To test the power of the medium, the client offered to send three cigars free of charge to every person who heard one of his programs and wrote to him. After the third week, WHAS received an urgent appeal: "Please stop immediately. Am using twenty girls to mail cigars and four days behind already. We cannot stand it. No one can stand it. Stop immediately."

Dry Humor

Jack Benny started on NBC in 1932 on a program sponsored by Canada Dry Ginger Ale. He would sometimes kid the sponsor by adapting one of the old Prohibition gags: "Her father drank everything in the United States and then went up North to drink Canada Dry."

Canada Dry lacked even the driest kind of humor and soon dropped the program.

A Suitable Case for Treatment

Henry Morgan had a half-hour comedy program on radio in the forties, which acquired a cult status among its followers. Morgan, whose real name was Harry van Ost, was an enfant terrible of broadcasting, riling management, sponsors, and any authority figures, while delighting his audiences with his irreverence.

Working as an announcer in Philadelphia, Morgan once included in his daily broadcast of missing persons the name of the local police chief and his boss at the radio station.

Once, a vice president of the Mutual Broadcasting System, which owned WOR, walked in on Morgan during a broadcast. "Who's that funny face that just came in?" the announcer exclaimed, and refused to continue until the man stopped his eavesdropping.

In a subsequent program, to liven things up, Morgan auctioned off the executive board of the Mutual Broadcasting System, fetching $83 for the lot; he threw in the premises and good will.

Morgan was once asked to deliver a weather report. "High winds, followed by high skirts, followed by me," he predicted.

Apologizing for one of his shows, Morgan explained: "This program was written while the author was under the influence of money."

Morgan vs. the Sponsors

Henry Morgan reserved his sharpest bites for the hands that fed him. He once demonstrated a safety razor by shaving in front of the studio audience. "We will continue," he announced, "as soon as we mop up the blood."

The sponsor was in his booth listening. The man groaned:
"He's slashing my throat with my own razor!"

Extolling the virtues of a brand of iodine, Morgan wound up by suggesting: "Try drinking a bottle for a broken arm."

As spokesman for a make of automobile, Morgan remarked: "Our cars are now rolling off the assembly line; as soon as we keep them on the assembly line, we'll start delivering them."

Morgan was fired by the makers of the Oh, Henry candy bar after a series of commercials in which Morgan would declare: "Yes, Oh Henry is a meal in itself." He then added: "But you eat three meals of Oh Henrys and your teeth will fall out."

The following week, he waxed lyrical over the virtues of the candy bar, only to wind up with the warning: "Feed your children enough Oh Henrys and they'll get sick and die."

A sense of humor was not the strong suit of Morgan's next sponsors. Life Savers canceled his program after Morgan tried to show that the public was being cheated because of the large hole in the middle of the peppermint drops. He announced that he would be soon marketing a supplement, Morgan's Minted Middles.

The Adler brothers manufactured elevator shoes, and they advertised on one of Henry Morgan's shows. They did not like it when they heard Morgan broadcast: "These fellows make shoes in ten different colors—five of which I would not be caught dead in."

The next day, the most senior brother, Jesse, known as Old Man Adler, came to see Henry Morgan, seeking an explanation for his comment. The comic said he was sorry for his error—in fact, he would be caught dead in those aforementioned shoes, but that was the only place he'd be caught.

Truth in Advertising

During a program sponsored by one of the tobacco companies, announcer Westbrook Van Voorhees had a coughing attack. He recovered sufficiently to explain:
"Guess I've been smoking too much!"

In the middle of a World Series broadcast in the fifties, sponsored by Gillette, an announcer asked Billy Southworth, who was then with the St. Louis Cardinals, whether he had used a Gillette shaver that morning. Southworth must have cut himself, because he replied unexpectedly:

"You know bloody well I did!"

The Truth Shall Make You Free

Don Sherwood, a famous disc jockey in San Francisco during the fifties and sixties, and a notorious drunk, once asked a friend's twelve-year-old son to do the commercial for Falstaff beer in his stead. Company executives were infuriated, cancelled the contract, and by the following day had stripped the studio of all their signs and paraphernalia used to advertise their product. The KGO-TV staff most regretted the end of free beer, but Don Sherwood faced the camera from behind his bare desk and said with a brave grin:

"There's something I've wanted to say on this show for a whole year, and I haven't been able to. Now I can. *Budweiser.*"

Biting the Hands

Sponsors—including their family, employees, or friends—frequently attended the broadcast sessions of the shows that they sponsored, both for entertainment value and to make sure the comedians did not go too far in mocking their products. They watched the show from a special booth, the radio-age equivalent of the royal box at a European Opera House.

During one of Fred Allen's warm-up sessions, a light flashed on in the sponsor's booth, revealing it to be empty.

"Ladies and gentlemen," Allen announced, "that booth is a device to belittle the comedian by showing him that the sponsor doesn't care enough for his program to attend it."

An usher quickly rushed to turn off the light in the booth.

"Ah," Allen remarked, "a boy who has the guts to turn off the lights without a memo from a vice president will go right to the top of the organization."

Arthur Godfrey, whom Fred Allen once described as "Peck's Bad Boy of Radio," continually got into trouble with sponsors. He would delight audiences by pronouncing Bayer's Aspirin as "bare ass prin."

Cletus Elwood (Boots) Poffenberger, the baseball player, was asked during an interview on a program sponsored by a cereal company:

"Now tell us, Boots, what is your favorite breakfast, taken with cream, sugar, and some sort of fruit?" Poffenberger looked the announcer straight in the eye, and told him the truth:

"Ham, eggs, and a couple of bottles of beer."

And another baseball great, Lou Gehrig, admitted, while interviewed on the Grape Nuts program, that his favorite breakfast food was Wheaties.

One time when his radio show was sponsored by a toothpaste company, Bob Hope was asked about the most embarrassing moment in his career. "A man from Pepsodent," Hope recalled, "walked into my hotel room and found me washing my teeth with salt."

Unlucky Strike

W. C. Fields was appearing on a radio program sponsored by Lucky Strike cigarettes. The comedian had everybody in stitches with his stories about an imaginary son of his called Chester. Only toward the end did the host put together the first and last name of the younger Fields, which spelled a rival cigarette brand and the end of the comic's appearance on the program.

The Mouths of Babes

In 1936 Cecil B. DeMille, in semi-retirement from directing films, undertook to produce a weekly radio drama program, which he also introduced and signed off. *Lux Radio Theatre* was sponsored by Lever Brothers, the makers of soap. The prestigious drama attracted up to forty million listeners during its long run and made deMille personally more famous at the time than any of the epics by which he is remembered now. However, it failed to impress his grand-daughter Cecilia. The director overheard someone ask what her grandfather was doing.

"He is selling soap," she replied truthfully.

One evening, Cecilia did not want to say her prayers. Cecil knelt with Cecilia at her bed and did a rousing rendition of the Lord's Prayer. The child remained obstinately silent until the end: "Thine is the kingdom and the power and the glory . . ." when she cut in:

"This is Cecil B. DeMille saying good-night to you, from Hollywood!"

Whose Business Is It Anyway?

In the days when advertising agencies used to produce radio and television shows for their corporate sponsors, there was sometimes a great deal of tension between the artistic types and Madison Avenue. Ken Whelan was directing for Colgate-Palmolive *The Big Payoff*—a game show in the fifties that featured Bess Myerson, the former Miss America and future Commissioner of Consumer Affairs in New York City—when he was introduced to Brian Houston, who ran an advertising agency.

"Someone mentioned that you used to be in show business," Houston said by way of small talk.

"Well, yes . . . most television directors are."

"Without being personal," the adman cut him off, "I have to say that I've always resented the way you show people sneaked into the advertising business through the side door."

"Without being personal," the director shot back, "I have to say that I've always resented the way you advertising people sneaked into show business through the back door."

A Certain Smile

Ed Gardner used to work for an advertising agency, where he did not distinguish himself in any way. Years later he became a great radio star as Archie on *Duffy's Tavern*, and he was invited back to do a guest spot on one show produced by his former agency.

The actor decided to get even with Howard Kohl, the dour personnel manager of the agency. He agreed to appear on the radio program for one dollar less than the $1000 fee offered. In place of the dollar, he insisted on a smile from Howard Kohl. A contract was drawn accordingly, but when the time came for the broadcast, Kohl was nowhere in sight.

Ed Gardner refused to go on the air, until the producers frantically searched for Kohl, dragged him into the studio and forced him to smile at his former employee.

Commercial Break

An NBC executive in the 1950s decided to try a little experiment about the effect of beer advertising, before it was outlawed on television. He was watching a fight on his network until he heard the announcer sternly bid him to go to the icebox for that ice-cold bottle of beer. Gulping furiously he managed to stay abreast of the announcer, getting a new bottle every time he was told to.

The score in this experiment resulted in the losing fighter being kayoed in the tenth, while the executive was flat on his back at the end of the ninth.

Up the Drain

Before a more exact science of measuring the audience for radio had been devised, New York's commissioner of the City Water Department was puzzled by a new phenomenon. At certain times in the evening, water pressure suddenly fell as if everybody decided to turn on the water all at once. Then, after a couple of minutes, everything became normal, before the whole cycle was repeated again. After some investigation, inspectors realized that the cause was *Amos 'n' Andy*, the top radio show in the late 1920s, which had people glued to their sets until a commercial break, when everybody rushed to the bathroom at the same time.

The phenomenon, which would become commonplace with other popular radio and television shows, entered for a while the business lingo of broadcasting. Whenever an idea for a new program was suggested, executives would ask: "But is it a toilet-flusher?"

Ratings

Few things are more hotly debated and resented inside the American broadcasting industry than the various methods used to measure the size of audiences. In the 1930s the Crossley ratings competed with Hooperating, both of which were conducted by making random telephone calls. In 1947, Mr. Hooper himself appeared on the program *We, the People* to demonstrate how he put together the sample for his Broadcast Audience Measurement.

"What program are you listening to?" he asked a person he had just called at random.

"*Amos 'n' Andy*," the voice replied, and the studio audience roared.

Hooper received only a Hooperating of 13, compared to 30 for the popular *Amos 'n' Andy*.

Since Americans love gizmos, audience measurements became more complicated and supposedly scientific. The Neilsen organization installed little black boxes, called audimeters, in a small number of households. These were attached to TV receivers and read when the sets were switched on or off and when the channels were changed. But the machine could not tell who or how many people were watching. One zealous audience research company experimented with measuring the sweatiness of the viewers' palms, one of the methods used on lie-detectors. And just as in polygraph tests, a neutral or safe question is usually asked first to establish the norm, so this company showed an old *Mr. Magoo* cartoon, which became the unit of measurement. A pilot program that scored above 5.1 Magoos was considered viable.

Producer Bob Shanks relates how he and Merv Griffin were shown the graph for one of the pilots they had just made. Griffin was excited to see that the graph shot up sharply at one point.

"Quick," the singer said, "tell me what I did there, so I can do it some more."

The researcher checked the script and replied:

"That's where Florence Henderson punched you in the stomach."

Master of All That He Surveys

The television networks like to use audience surveys to support their choice of programming. Even though William Paley, the founding chairman of CBS, greatly disliked at first the situation comedy *All in the Family* for its vulgarity and bigotry, he became an enthusiastic convert once the show emerged as the vanguard of a successful new trend. TV critics began to argue whether the sitcom might be helping to reduce racial prejudice by making fun of it. CBS readily commissioned a survey to offer a scientific basis for proponents of the theory. Much to management's dismay the study came to the opposite conclusion, and Jack Schneider, president of the CBS Broadcast Group, went to his boss with it.

"If we release this," he gave Paley the report, "we'll have to cancel the show."

"Destroy the study," the chairman counseled.

A few years later, *Beacon Hill* became an expensive flop on CBS. The upscale show, modeled on the British series *Upstairs Downstairs*, had been a pet project with William Paley and he ordered a study to determine the cause of its failure. One conclusion in the survey stated that the American public simply did not believe that a single household would have so many servants.

Catcalls

The American television networks conduct extensive surveys and tests both before they create a program and after it has been aired. Deborah Pietruska, working for the NBC Program Research Department, recalls sifting through a pile of questionnaires, when her eyes fell upon this comment from a female viewer:

"The show was so disgusting, the cat left the room."

Morris the Cat won great name recognition in the mid-eighties from a series of memorable catfood commercials. He was drafted as Democratic candidate to enter the 1988 presidential race.

"The world is going to the dogs," one of Morris's handlers announced to the press: "America needs a president with courage, but one who won't pussyfoot around the issues of peace—a president who, when adversity arises, will always land on his feet."

The Secret of Our Success

Actress Farrah Fawcett commented on the early ratings of her first successful series, *Charlie's Angels:*

"When the show was number three, I thought it was our acting. When we got to be number one, I decided it could only be because none of us wears a bra."

Before becoming famous as one of *Charlie's Angels*, Farrah Fawcett was already well known for more than a hundred commercials she had done for toothpaste and skin cream, among many other products. But since people did not know her name, fans would stop her or call out to her: "Hey, Ultra Bright!" or "Hi there, creamy!"

I Dream of Carole

Although *I Love Lucy* quickly became the top-rated comedy show on television, it did not look like a sure thing in the winter of 1950–51, when Lucille Ball and her husband Desi Arnaz tried to sell the pilot to CBS. They founded Desilu Productions to create their own television show with $16,000, half of it borrowed. It was estimated that they gambled away half a million dollars they might have earned in film, radio, and band bookings.

At the time Hollywood was still trying to ignore the threat posed by television to its box office, and many studios practically forbade their contract players to appear on the upstart medium. Snobbery would keep others from the small screen in a comic repetition of history: some fifty years before, many actors of the legitimate stage boycotted the newfangled cinema with the same feeling of superiority. Lucille Ball had been a Goldwyn Girl in the 1930s, and she had been working steadily in movies, even though major stardom had eluded her. So what made her decide to take the plunge into television?

She always used to say that one reason was to save her marriage, by working more closely with her husband. The formula worked for a while, even though by the last episode in 1959, their stormy marriage was over. The other incident that influenced Lucy was a dream, in which she saw her friend Carole Lombard, whose plane had crashed in 1942, appear.

"She was wearing a very smart suit," Ball would recall, "and she said, 'Honey, go ahead. Take a chance. Give it a whirl.'"

I'll Never Go Hungry Again

Desilu Productions became a phenomenon. In less than a decade, the company that began with $16,000 owned several film studios and more sound stages than mighty Metro-Goldwyn-Mayer. The sweetest stroke of all was the purchase from RKO-Pathé of the old studio where David O. Selznick had made *Gone With the Wind*. As a young contract player at RKO, Lucille Ball was sent along with hundreds of hopefuls to audition for the part of Scarlett O'Hara.

The story goes that she lost her way to Culver City, got caught in a downpour, and arrived for the audition disheveled and wet. Selznick was late for the appointment, and the actress knelt in front of the blazing fireplace in his office, trying to dry out. When the producer finally arrived, they read a scene a together, before he politely dismissed her. Only then did Miss Ball realize that she had done the entire audition on her knees.

After buying the studio, Lucy—whether out of nostalgia or revenge—claimed the former office of David O. Selznick for herself.

They Stayed Home in Droves

When television began to keep audiences away from movie theatres, Samuel Goldwyn was one of the early Hollywood moguls to see the trend. "Who wants to go out and see a bad movie," he declared, "when they can stay at home and see it free on television?"

Progress

After Ed Wynn made a comeback in television, the old vaudevillian who began performing at fifteen remarked: "In the old days you could see me for fifty cents. And look how I've progressed—now you can see me for nothing."

Hall of Fame

Ed Wynn was talking to one of the writers of his television show, Seaman Jacobs.

"I just don't understand it, Sy. They have the Baseball Hall of Fame. They build big monuments to the presidents. They put famous scientists on stamps. But there's no place that honors the great comedians."

"Sure there is, Ed," Jacobs replied. "It's called the Bank of America."

Where the Chips Fell

Bert Lahr, the great comic, could not get much work on television; as one bright young Hollywood producer said about the Cowardly Lion in *The Wizard of Oz*, the most watched film in history: "No one knows who Bert Lahr is."

Toward the end of his life Lahr was appearing on television commercials for the Frito-Lay company. He told Carroll Carroll, his producer at the J. Walter Thompson advertising firm:

"Ain't it ironic? After all the things I've done in show business, I gotta be kept alive by a potato chip?"

Pro Bono

Richard Rodgers, the musical composer, was working on a television special when he thought of the idea of asking Milton Berle to do a cameo. He called the comic to ask how much he would charge.

"Dick, I'm so flattered that you want me," Mr. Television replied, "that money will be the very last thing I think of—before I get to sleep tonight."

Hardware vs. Software

Before *Your Show of Shows*, Sid Caesar began his comic career on television with the weekly program, *The Admiral Broadway Revue*. It was canceled by the sponsor, a manufacturer of television sets, for being too successful. Ross Siragusa, president of the Admiral Corporation, summoned Caesar to Chicago to explain that orders for television sets shot up from five hundred to five thousand a week within the first three weeks of the show. Three months later, orders had climbed to ten thousand sets. The company's resources were so strained that it had to choose between sponsoring the program or investing that money into making more sets.

"Do you mean to tell me," Caesar protested, "that the show is being canceled for bringing in too much business?"

Mr. Siragusa was all smiles:

"I'm so glad you do understand, Mr. Caesar."

Standing Up to Caesar

The Internal Revenue Service objected to Dinah Shore deducting the cost of expensive gowns she wore in her television and stage appearances. After lengthy legal arguments, the singer demonstrated to the tax collectors that she could not possibly use those dresses at home or in everyday life. The IRS was convinced and issued what became known as the "Dinah Shore" ruling: a gown was deductible if it was too tight for the performer to sit down in it.

It's a Bust

Making entertainment programs involves thousands of details, of which the audience knows mostly nothing. Just the legal problems that have to be solved against the most severe deadlines—and usually without litigation—would astonish most people who rightly assume that the legal system was created to achieve just the opposite results. Michael Donaldson, one of Hollywood's finest entertainment attorneys, who spends much of his day with resolving such matters, describes a typical incident that involved a television film directed by his client, Michael Landon.

I remember once trying to get permission to use the likeness of a deceased star. Technically, under the law, it probably wasn't necessary and I explained that to Michael, but the bust was a plot point in the script about a blind sculptor, and I didn't want the bust to show up on a television screen if it might upset relatives of the star. When I contacted the attorney for the estate, he didn't understand at all. He threw out a large demand. I tried hard to explain to him both the law, which he knew well, and the reason we wanted formal permission: sensitivity toward his clients. He still didn't get it: it was all about money.

When I explained the situation to Michael, he authorized a modest but fair payment and asked me to close a deal on that basis. It was late in the day, and the other lawyer thought he had us because the relevant scenes had to be shot the next day. He not only held firm but intimated that the price might go up the next day. Wrong move.

The following morning I went to the set, where work had already begun. I explained the situation to Landon and advised him strenuously against meeting the lawyer's high demand. Michael looked me right in the eye and said:

"I don't care about that attorney. I do care about the family. I'm not going to use this without their knowledge."

He strode off. He got behind the camera and looked at the scene that was about to be shot.

"Alright, in this scene the bust is completed," he told the actor playing the blind sculptor. "I'm going to change the camera angle. We will come in from the back of the bust and shoot your face, reflecting the pride of your accomplishment. I want you to touch and feel it all around and slowly turn angry and push it off the table. The camera will pull back, catch it falling on the floor and breaking into pieces."

Everyone was surprised. This was a major change in the script. Michael Landon shot the scene, got it in one take, and made a few minor adjustments to the story line. He had been fair and firm. He was often generous. But never, ever, to be messed with.

We never heard again from the attorney who had blown the deal.

How Tragedies End

After commercial television was introduced in Britain in the mid-fifties, there were many complaints against advertising that interrupted programs. In a famous instance, a live broadcast of *Hamlet* ran too long, and the ending was replaced with an ad extolling the virtues of a certain brand of orange drink. Lew Grade, who as head of the independent television system was particularly sensitive to backlash from such accidents, called up his station in a rage:

"What the devil happened?"

"Oh," said a bland voice at the other end, "they all died in the end."

Making the Grade

Lord Grade, sometimes called "The Last of the Great Showmen," began as a music-hall performer with his brothers, Leslie and Bernie. Robert Morley, in his eulogy of Leslie Grade, said in 1979: "Sometimes I'm asked when I'm abroad if England isn't a rather stick-in-the-mud country, where there's no opportunity for people. My goodness! A country that's produced a family like the Grades? I mean, where else in the world could a Charleston team end up in the House of Lords?"

Lew Grade worked as an agent for a while. He was once watching a variety act which reminded him of what he used to do on stage with his brother Bernie. Afterward he went backstage and told the actors how much he liked them.

"It was marvellous!" he slapped them on the back and got down to business. "How much are you getting?"

"Twenty-five pounds a week, Mr. Grade."

"That's ridiculous, I can get you forty," said Grade. "Who's your agent?"

"You are, Mr. Grade."

Principles

In the 1960s the Hungarian-born British humorist George Mikes was collaborating on a *Book of Snobs* with the thirteenth Duke of Bedford, who had scandalized the English peerage not only by commercializing his country estate but by flaunting his talent and enjoyment at making money. French television expressed an interest in turning the book into a two-hour show, and dispatched a high-level executive to ask if parts of it could be filmed at Woburn Abbey, at His Grace's ancestral seat.

In his memoirs, *How to Be Seventy*, Mikes recalls being invited by the Duke to Woburn to hear about the proposition. The Frenchman was overwhelmed by the magnificent surroundings (22 Canalettos hung in the dining room alone), but managed to outline the program, which was to be written by Mikes, and narrated by the Duke. The authors liked the idea, but His Grace only had one question:

"How much do you pay?"

The man from Paris became deeply embarrassed and quickly explained that such matters would be discussed later by someone below him.

"Will you agree to do it, Your Grace?" he asked anxiously, to which the Duke, looking and sounding like "the assistant bank manager of a suburban branch," returned his question:

"How much do you pay?"

The French producer, frustrated and sweating, tried to explain once more:

"I am a producer and an artist. I should like to discuss the matter in principle. Does Your Grace agree in principle?"

"If you pay enough, I agree," replied the thirteenth Duke of Bedford, "if you don't pay enough—I don't agree. And I have no other principles."

Let's Kill All the Clients

Until relatively recently, professional ethics, if not the law, prevented American lawyers from advertising. Now that the inhibition has been shed, some attorneys have gone far out on a limb. Rob Moore in Orlando, Florida, reported that he heard one lawyer, specializing in personal injury cases, begin his television pitch: "If you or any member of your family has been killed . . ."

Tight Fists

When the American networks canceled all regular programming and commercials in the days following President Kennedy's assassination, ABC was in the weakest position to absorb the costs. Before satellites, the networks had to rent a special cable through Ma Bell to keep the phone lines open between their Dallas affiliates and their news headquarters in New York. That cost alone came to $1.15 per minute, per mile.

Mike Boland, who worked for ABC's legendarily tight-fisted controller Si Siegel, kept moaning to the news division about the mounting transmission costs until somebody, overwrought with fatigue and emotion, had enough:

"I'll tell you what, Mike," he said. "The next time they assassinate a president, we'll have 'em do it in Brooklyn."

Don Carswell, longtime NBC vice president in charge of financial planning, made sure his office projected his style of doing business. There hung a picture of a huge, brutish Scottish laird wielding a broadsword, with a caption underneath: "The Spirit of Compromise."

And any suppliant reaching his desk would face a plaque which said: "The answer is No."

Expensive Account

Joe Aceti, when he was a producer on *Wide World of Sports*, submitted his expenses incurred while shooting the Walker Cup Amateur Golf Tournament in England. An ABC business manager noticed one item which read: "Tree trimming—$50." Clearing foliage to make sure the camera has an unobstructed view is a normal expense, but the manager happened to have watched the tournament, and he summoned Aceti.

"What do you mean by tree-trimming?" he asked. "I saw the opening pan of the whole golf course and couldn't see a shrub over two inches high." Without a moment's hesitation, Aceti responded:

"Boy, that guy really did a good job!"

The Bottom Line

Bill Leonard took over CBS News in 1975, and had a meeting with his new boss, Kidder Meade, at "Black Rock," the CBS headquarters building in New York. During their discussion, Meade's secretary came in and handed him a small, folded piece of paper. Meade read it with satisfaction and then handed the paper to Leonard. It had only a number written on it: 38½.

"That's the price of CBS stock at noon," Meade said. "That's the bottom line. You're not in the news business any more."

Invasion of the Flesh-Peddlers

Trying to stem the rising costs of network news in 1983, Ed Joyce, then president of CBS News, laid the blame for skyrocketing salaries on super-agent Richard Leibner. He accused the agent of forcing the networks into competitive bidding, which resulted in many veteran correspondents choosing to leave CBS. The attack on "flesh-peddlers"—an old Hollywood term employed by Joyce—caused a storm of controversy and brought him the enmity of the star reporters he was trying to keep. One of them, Paris correspondent Don Kladstrup, wired his protest, which contained the line:

"And to think I was worried that Beirut was the only place I had to watch my back."

Major Major

In television, writers and actors spend an inordinate amount of time talking about agents who are the real power brokers in Hollywood. One well-known agent liked to drive home his clout by adding an extra little touch, as writer Linda Elstad tells the story.

Leonard Hanser was head of the Major Talent Agency, so called because he had been a major in the army. Anyway, Hanser was a very short guy with a very big ego, and he was a general when it came to negotiating or commanding his clients. One of them was my friend Bill Blinn, for whom Hanser was negotiating a very big deal for a major series. Each day, hour by hour, Bill was waiting nervously by the phone as Hanser relayed the latest terms which the writer would have gladly accepted, only to hear from his agent that the terms were not quite good enough yet and that he intended to squeeze the producers for more. This went on, while Bill was getting desperate and close to a nervous breakdown. Finally, after a lengthy wait, Leonard phoned Bill.

"Has it happened?" Bill asked impatiently.

"I can't tell you yet," said the agent calmly. "First I want you to look out your front window, and come back to the phone and tell me what you see."

So Bill went to the window, where he saw a delivery truck at his door, with people unloading some large boxes. Bill returned to the phone and told his agent what he saw. Hanser said:

"You've got the deal."

They were delivering a great big TV set and a VCR, which was something of an expensive novelty in those days. Major Hanser, after he negotiated every dot on every *i*, further stipulated that there would be no deal until the goodies were delivered—at once—to his client's front door. It demonstrated that he had the power beyond the letter of the contract to demand whatever he wanted. There were stories about similar deals, including a racehorse once, which—unlike in *The Godfather*—came with the head and body attached.

One-downmanship

In 1982 Group W and ABC launched a twenty-four-hour cable news service to compete with Ted Turner's Cable News Network. Satellite News Channels offered its services for free, as opposed to CNN, which charged cable operators. Turner was forced to drop these charges, started yet another twenty-four-hour network (CNN2, which became Headline News), and offered other inducements to the operators. All this brought him to the brink of financial ruin, but his competitors blinked first. They offered SNC to Turner, who bought it, only to shut it down.

Some time later, Herb Granath, who ran ABC Video Enterprises, ran into Ted Turner at a party. The latter confessed that if the competition had lasted another six months, "I would have been tapped out," but that the bluff worked.

"We were going to fold SNC anyway," Granath tried to trump the Atlanta tycoon, "and then you came along and gave us $25 million so *you* could fold it."

Ted Turner was losing a million dollars a month on CNN when he rushed into setting up a second twenty-four-hour network, giving his staff only four months to get the new headline service on the air. Ted Kavanau, given charge of CNN2, told Turner that he could get it on the air in half the time.

"Why are you in such a hurry," the media mogul quipped, "for me to lose another million bucks a month?"

Real News

Soon after they reported on the bombing of Baghdad in the first phase of the Persian Gulf War, CNN correspondents Bernard Shaw and John Holliman left on the dangerous overland route to Amman, Jordan. They were out of contact for twelve hours. When they arrived at the Jordanian border, their boss, Ed Turner, greeted them with the news that he would be docking their pay for those twelve hours.

Biblical Parallel

Ted Turner was just launching Cable News Network when he confided to David Frost that he was bound to make some mistakes in finding suitable staff:

"After all, Jesus Christ only had to make twelve appointments and one of them was a bummer!"

Unconventional Confession

Sister Carol Anne O'Marie is a nun in Oakland, California, who writes mystery novels about an elderly nun playing detective. According to Leigh Weiners of the *San Jose Mercury*, Sister O'Marie was recently approached by a Hollywood company to turn her novels into a television series.

She was told that it would help dramatically if the central character were younger, had a drinking problem, and perhaps had an illicit love affair before she donned the habit. When the author declined to contemplate such changes, the television producer tried the ultimate argument:

"You're turning down a chance, sister, to make a lot of money."

"What would I do with it?" replied O'Marie, who had taken a vow of poverty: "I'm not going to live in a nicer convent."

They Are Different from You and Me

Ben Stein, former media critic, novelist, and teacher who became a successful actor in Hollywood, made a study or how the television industry worked in the late seventies. He met many people who were making as much money in a week as a teacher might in a year, and yet they did not consider themselves rich. "All, including the most highly paid," Stein observed in *The View from Sunset Boulevard*, "were acutely conscious of the fact that real wealth in America came from manufacturing or banking or something they were not doing." Stein asked a producer who was making a salary of more than $250,000 a year how to make money in Hollywood.

"Have a father who manufactures shoes," the man said.

CHAPTER 14

Medium Rare

Toward World Peace

After installing a radio, Gertrude Stein reported: "I never listened to one before. There is a deplorable amount of music going on in the world; if they would suppress most of it perhaps the world would be more peaceful."

Golden Silence

George S. Kaufman spent his last years bedridden, listening to radio. He was annoyed by a musical request program which played a few minutes of the music, before fading out for a commercial. He could not get up and turn off the radio, so he called up the announcer, who was delighted to have the famous writer on his show.

"What's your wish?" he asked, "I'll play it right away."

"My request is for five minutes of silence," Kaufman said and hung up.

Canned Silence

Commenting on the strange and sometimes sick topics that studio audiences for television sitcoms would laugh at, Dr. Harvey Jacobs, a psychiatrist, observed in the early seventies:

"Not all of it is laughter. I would estimate that only about twenty percent of the audience in the average sitcom studio are inveterate guffawers. A lot of laughing is actually repressed coughing. When they leave the studio, eighty percent of the Los Angeles studio audience goes into an immediate depression."

Newsday columnist Marvin Kitman observed that what sitcoms really needed was "canned silence," but that it would be an "anathema to TV, with its high standards for honesty."

Dr. Jacobs's recipe for achieving silence with a live audience would be "to tape the shows in a Trappist monastery."

Paternity

During an argument on *The Big Show* Tallulah Bankhead told playwright John Van Druten that she thought radio was the mother of television.

"And who's the father?" asked Van Druten.

"Television has no father," Miss Bankhead replied.

The Medium With No Message

It was also on *The Big Show* that Fred Allen delivered a classic line, supposed to have been written by Ace Goodman: "I've decided why they call television a medium. It's because nothing in it is well done."

Groucho Marx elaborated on the dictum: "They call TV a medium, because so little of it is rare or well done."

"I didn't make it in television," Fred Allen explained once, "because of ill health. I made people sick."

Fred Allen was visiting Bob Hope's suite at the Algonquin Hotel when he noticed a large arrangement of flowers and fruits on top of the TV set.

"You know," Allen remarked wanly, "that's the best thing I've seen on television yet."

Fred Allen attacked the new medium whenever he could. One time, he called television "a coffin with a window," and explained: "Vaudeville is dead, and television is the box they buried it in."

On another occasion he observed: "Television is called a medium, because it puts you in touch with so much that has expired."

Fred Allen, the quintessential radio comic, feared television. At first he hoped it would not last: "There are still several states in the union," he once told his radio audience, "where they think Television is a city in Israel."

When movies began showing up on TV, Allen swore he saw a Western "so old that the cowboy was riding a dinosaur."

"The only performers who will last in this medium," he once predicted, "will be the pointers. Pointers never do anything themselves. They merely stand center stage, point to another performer, and announce: 'See that fellow? He's going to do the darndest trick you ever saw!' Then the other fellow comes out and does the trick. A week later, the pointer is back gesturing at somebody else, but the fellow who did the great trick has already given his all and is out in the cold. TV can eliminate pointers if times get tough enough. They can teach dogs to do the same routine simply by smearing meat on the actors."

Sic Transit

Goodman Ace was one of the great radio comics who did not feel comfortable with television, even though he would become one of its highest paid writers. As the new medium began to overwhelm radio, Goodman mournfully told a friend:

"If an atomic bomb ever falls on New York City, I want to be on the eighteenth floor of the CBS building, because there is no radio activity there at all."

Asked about the difference between writing for radio and writing for television, Goodman Ace said: "Well, in television you only write halfway across the page. It doesn't mean much to anyone else, but it does double a writer's overhead."

Medium Cool

In the summer of 1969, just before the first manned landing on the moon, William O. Johnson, Jr., an editor for *Sports Illustrated*, was working on an issue about the impact of television on sports. He decided to interview Marshall McLuhan, the eccentric Canadian professor of English, who gave us buzzwords and slogans such as "the global village," "the medium is the message," and "hot and cool media."

Johnson reached McLuhan directly on his first try, but the guru told him that he was sick of giving away his ideas for nothing or having them stolen by Madison Avenue. When the journalist explained that he was simply after a comment, McLuhan told him bluntly to use whatever he has already published, but that he would fail anyway with writing the story:

"You couldn't possibly understand it. You have only concepts. You don't have perceptions."

Johnson pressed him, but McLuhan then veered off at a McLuhanesque tangent:

"The moonwalk is going to have an incredible impact on sport. An immense impact."

Johnson wanted to know what the professor meant.

"Of course, I won't tell you," McLuhan replied. "But don't forget, the moonwalk is going to have an impact on sport for all time. Don't forget. You heard it here first."

Then he hung up.

Stranger Than Fiction

Perhaps the most penetrating and enduring critique of the mass media is the one found in *Nineteen Eighty-Four*. George Orwell began writing his utopian novel just after leaving the BBC, where he worked as a Talks producer for the Indian section of the Empire Service. Malcolm Muggeridge claims that Orwell once told him that he had conceived the idea for the Ministry of Truth from his experiences at Broadcasting House. Martin Esslin identifies 200 Oxford Street, a department store that the BBC requisitioned during the war, as the more exact parallel for the honeycomb of small cubicles where Winston Smith worked amidst the murmur of voices dictating. The most dreary description was reserved for the canteen, in the basement of the building.

"The cups were chipped, the food was foul," Esslin recalled. "The younger people had been called up for military service, so the average age of those left behind was high, and they were anything but attractive; the elderly secretaries might well have been members of the Anti-Sex League. There was also a shortage of canteen and cleaning staff, so the table that one finally found empty was invariably covered with dirty dishes and pools of slops, exactly as in Orwell's description.

During the day and evening the canteen was crowded, but work in the Empire Service went on throughout the night, as broadcasts were being beamed to parts of the world in widely different time zones. At two or three o'clock in the morning, the badly cleaned, stuffy, blacked-out, windowless building was at its most depressing; an atmosphere of hopelessness and futility oozed from the dimly lit corridors and the empty, echoing canteen. Anyone who had experienced these night shifts instantly recognizes their mood in the pages of Orwell's novel."

Management Training

Sir Tyrone Guthrie, the great stage director, worked early in his career as a radio producer for the BBC in Belfast. (Indeed, he had written some of the earliest radio dramas on the BBC.) In the 1960s, when he was world-famous, the BBC invited him back to give an after-dinner address to one of the management training seminars that were regularly held at a posh country house. Martin Esslin, head of Radio Drama, was supposed to greet Guthrie. He recalls that the man who had organized the seminar was a highly officious executive, who had conducted a lengthy correspondence with Guthrie over every detail of the planned talk.

At 7:45 on the given evening there was no sign of the guest speaker, and when he finally turned up at 7:58, he brushed by the organizer and pulled Esslin aside.

"Martin," Guthrie asked, "what am I supposed to talk about?"

"About the BBC and the Arts," Esslin whispered back.

After being introduced at some length by the officious organizer, Guthrie looked down from his great height at the gathering of present and future managers, and said:

"Ladies and gentlemen, if any artist asked me whether he should have anything to do with the BBC, I'd tell him: 'Fuck off!'"

In the silence that followed, the great man sat down. Then ensued, Esslin recalls, one of the liveliest and most enjoyable discussions on record.

No Bull

From its early days, the BBC provided training in the techniques of enunciation both for announcers and actors. One of the experts in spoken English went by the name of Commander E. J. King-Bull, who once introduced himself to an actress simply as King-Bull.

"And I'm Queen Cow," she replied.

Class Struggle

Before he became famous as the premier quizmaster of *The Brains Trust*, Gilbert Harding hosted *Round Britain Quiz*, for which he had to travel all over the British Isles. His first trip in 1947 took him to Glasgow to meet his panel. Harding was worried that a radio audience might not be able to distinguish between the accents of Sir James Fergusson, an old Etonian and baronet, and Jack House, a scrappy Glasgow journalist.

"I don't think they'll find any difficulty," House told Harding. "James speaks with a public school accent, while mine is a public house one."

English As a Second Language

Broadcasting on a national basis brought out the differences in regional accents and dialects, causing efforts to eliminate these variations. In time, the so-called "mid-Atlantic" *pronounciation emerged, which makes American and Canadian announcers sound halfway British. In* 1931, *Radio Weekly, published in Chicago, claimed that CBS and NBC were forcing their announcers to use English "as she is spoke in England." H. L. Mencken called NBC. "What we try to get," a Mr. Walter C. Stone reassured him, "is decent American pronounciation, affected as little as possible by localisms." In England, the problem of regional dialect has been compounded by class differences, as Sydney Newman, one of the pioneers of British television drama, told Denis Norden for his book on early television in Britain.*

No Trams to Lime Street, by Alun Owen, was the first play we did that really made TV history in Britain. It had a marvelous cast—Billie Whitelaw, Jack Hedley, Alfred Lynch and Tom Bell.

It took place in Liverpool. It went out on the air and the following day in my office at Teddington ten minutes after I read the rave reviews in the press, my phone rang. It was my managing director, Howard Thomas, who said, "Sydney, will you please come over to my office right away?"

I figured, great, he's going to pat me on the back and say it was a lovely show last night. So I went over but Howard was looking somewhat glum. Seated next to him was his boss, the Group Managing Director of Associated British Picture Corporation, C. J. Latta, an American. I soon guessed that Latta had just torn a strip off him, because he couldn't understand "the language talked last night." Well, I being a Canadian and him being an American, I talk one side of the mouth and he speaks on the other side, so I said, "C. J., they were talking with a Liverpool accent and, after all, we're North Americans. It might be tough for us, but everybody in Britain got it."

"Don't give me that," he said. "My chauffeur didn't understand it, and he's English. You tell those darn actors next time to speak fucking English."

Upstairs, Upstairs

England is famous for its class distinctions, which are still alive and well. Jean Marsh, one of the creators of the British series *Upstairs Downstairs*, could have had her pick of any role. When she chose to play Rose, the upstairs maid, her mother was disappointed.

"I thought you'd like it, mom," the actress said, "since you're working-class."

"No, we are upper-working-class," her mother corrected Marsh.

Going His Way

During rehearsals for one of Bing Crosby's early radio shows, a CBS director tried to correct the star's pronounciation of "tomayto."

"You should say 'tomah-to,'" the director told him. "It's got more class." The singer laughed, but did not go along:

"Listen, you Bronx peasant, I ain't sayin' 'tomahto' until I got a million dollars."

Of course, by the time Crosby got his millions he could afford to change his mind, rather than have to give up his pronounciation.

Better Than It Sounds

When I was in Britain recently promoting one of my books, Siobhan Synnot of Radio Scotland told of attending a program meeting at her station in Glasgow a few years back. Her boss began discussing a list of possible stories, including psychopaths in the city. The program director spoke of the prevalence of psychopaths in many areas of town, and urged his staff to get to know them firsthand. He went on at some length and in greater detail, as Synnot exchanged uncomfortable looks with some of her female colleagues at undertaking such an assignment. They were much relieved when their boss finally wrote out the list on a blackboard, beginning with the word CYCLE-PATHS.

Great Expectations

Robert MacNeil was being interviewed on ABC's *Good Morning America* about his book of reminiscences, *Wordstruck.* In it the co-host of the popular PBS news program, *The MacNeil-Lehrer NewsHour,* describes how his mother taught him to enunciate his vowels, when he was still a little boy growing up in Canada.

"Perhaps your mother already had in mind for you a career in broadcasting?" the interviewer suggested.

"Oh, no," MacNeil replied, "I think she had hoped for something better than that."

Unforgettable You

Back in the sixties, British comedy writer Keith Waterhouse was in a London theatrical club having a drink with a television director who had the habit, not uncommon in show business, of flattering everybody indiscriminately. An actor passed by, and the director congratulated him at once on his performance in

Rosencrantz and Guilderstern Are Dead. The actor was grateful, though he pointed out that he had not been in Tom Stoppard's play.

"*Othello*, then?" the director went on. "Did I see you in Olivier's *Othello?*"

The actor replied that he had never worked at the National Theatre.

"But I know I've seen you in something—I remember being very impressed. Very impressed, indeed."

The actor searched his mind but confessed that he had only acted in a radio play for the whole of the past year; in fact, he was working as a clerk at Harrods, in the groceries department.

"That's where I've seen you," the television director sighed with relief. "And you were bloody marvelous!"

Experience

Some stage actors had a difficult transition to make from theatre to radio drama, because they were used to performing in front of people. When producer Bill Spier was casting the series *Suspense*, he usually asked actors with a long theatre resumé:

"Have you had any experience playing without audiences?"

"Certainly," one of them replied. "I used to play Shakespeare on the road."

Party Time

In 1976 NBC inaugurated a new program called *The Big Event*, consisting mainly of the screening of well-known movies. But one week in September the network experimented with something called *The Big Party*, which it described as a live event with celebrities in New York City "to celebrate the start of a new season in motion pictures, the theatrical arts and sports."

During one segment Lauren Bacall interviewed Dustin Hoffman about his new movie with Laurence Olivier, *Marathon Man*.

"Many in our business," Miss Bacall said, "think Olivier is the greatest actor in the world."

"He must have the greatest taste of any actor," said Hoffman. "That's why he's not here tonight."

Let's Have Lunch

Composer David Shire and lyricist Richard Maltby, who have written a number of Broadway and off-Broadway shows together, began their collaboration while students at Yale. Their early success led to an invitation to work in Hollywood.

The two drove West, and on their very first day there an acquaintance introduced them to Marc Connelly. The venerable author of *Green Pastures* and other plays, who happened to be a Yale graduate, immediately invited the two for dinner.

Richard Maltby was quite overawed to be in the presence of Connelly, and confessed to him that seeing *Green Pastures* had been a seminal experience in his life. Conversation turned to a recent television version of the play. Maltby went on to express his opinion that how this piece of trash had failed to do justice to the original. Who, he wondered, could have perpetrated this thoroughly inferior version.

"I did," said the writer in a matter-of-fact voice, though his demeanor changed perceptibly. He asked the waiter for the check, and quickly walked out on the promising pair from Yale.

Where the Stories Come From

Garrison Keillor, creator of *A Prairie Home Companion*, a show that attracted millions of listeners to National Public Radio in the eighties, is one of the great storytellers of our time. Asked where he gets the characters and stories of his imaginary community of Lake Wobegon, Keillor claims everything comes from observation.

"I don't believe I invent anything," he tells interviewers. "I get in my car and drive a little bit west of Minneapolis. You don't have to drive far to be out beyond the life of the arts and, you know, pasta, to get where my stuff is set. Hutchinson, Hamel, Rogers, Watertown—they're all within easy driving distance. There are taverns in those towns that are really the social centers, where not everybody is drinking to get drunk. You don't want to dress funny; you don't want to look at anybody. You just want to sit there with your head down holding your beer. And a lot of their material is a heckuva lot funnier than what I do on Saturday night. But they don't operate with the same restrictions of good taste."

On Hold

One of the more offbeat characters on the American airwaves is the monologuist Joe Frank, who tells his bizarre and sometimes depressing stories in a unique voice over National Public Radio. A few years ago, Frank came out to California on a grant to produce his shows at KCRW in Santa Monica. He was very poor and staying in a rundown hotel. My friend Brad Schreiber, a writer, was already a fan of Joe Frank, and he happened to be working down at the studio when one of the monologues was on the air. It was late, with hardly anyone else around, and he happened to notice that one of the blinking buttons on the phone indicated that somebody was on hold. Schreiber picked up the phone and said:

"KCRW—can I help you?"

From the other end came the somber and unmistakable voice of Joe Frank: "No."

"Oh, who are you holding for?"

"I'm not holding for anyone."

"Is someone else helping you?"

"No," the voice said.

"Is there someone else you need to talk to?" Schreiber persisted.

"No, just put me back on hold."

Schreiber did so but told his co-producer Jon Cavanaugh:

"I'm pretty sure that Joe Frank is holding on this line. Why would he ask me to put him on hold and not want to talk to anyone?"

Cavanaugh knew Joe Frank and had an explanation:

"Joe's hard up, staying at a fleabag of a hotel down in Venice and probably doesn't have a radio in his room. When this station puts you on hold, you get whatever program is on the air. So this is the only way that Joe Frank can listen to himself."

Next We Will Try It on Humans

Maurice Dreicer, an eccentric millionaire, decided that radio might do something for those unfortunate dogs in New York City who could not get a good night's sleep. He once produced a series of midnight programs on WOR which mixed poetry with the sound of beefsteak bones being crunched. Media critic Ben Gross questioned him whether the sound effects might not actually keep the canine audience awake.

"Ah, my boy, you are wrong," said the producer. "First, my poetry puts 'em to sleep; then, my bones bring 'em dreams of luscious steaks!"

Loose Screws

Time magazine in the fifties quoted this passage from *The TV Technician's Handbook on Customer Relations:*

"There will be occasions when in order to make time you may inadvertently have a few small screws left over after repairing a television set. Solution: either put them in your pocket when the customer is not looking or mingle them with other items of the same type you may have in your tool kit."

In the same period, *Collier's* magazine reprinted this sign in a California TV store:

"TVs sold, installed and serviced here. Not responsible for summer programs."

What to Do for an Encore?

Programming executives make their decisions about which series to buy based on a sample episode, known as the pilot. Hundreds are made each year, few are chosen. Comedian Alan King once observed:

"I've lost more pilots than the Luftwaffe."

Marvin Kitman told the story in *Newsday* how CBS got one of its most successful series, *Hogan's Heroes*. The pilot was first offered to NBC, and during the screening hardened program executives were seen rolling in the aisle. Afterward they huddled to discuss the project. They decided to pass, after somebody raised the question: "If the pilot is this good, how could they sustain it in the second week?"

Different Strokes

In the late fifties, Leonard Goldenson, the founding chairman of the ABC network, went on a trip to the Middle East hoping to develop an appetite for television, which had not yet been introduced into the region. He failed with two customers for very different reasons.

He first met with the Emir of Kuwait to discuss setting up a television station in his country, but the only thing the Sheik wanted from Goldenson was a television set—made of solid gold. The American businessman could not make him understand that he was in the business of producing programs, without which even a gold television set was useless.

On the same tour, Goldenson went on to Israel, where he told Prime Minister David Ben-Gurion that he had the financial backing of British industrialist Sir Isaac Wolfson and of the Rothschilds to establish an Israeli television center. Ben-Gurion resisted all arguments, including the one that some Israelis in the northern part of the country had already bought sets to watch signals from Beirut.

"I won't allow my people to watch television," Ben-Gurion declared firmly. "They've got to work!"

Feeling Blue

The ABC network was formed in the forties when the Federal Communications Commission and the U.S. Supreme Court ruled that NBC could not run both its Red and its Blue networks. Indeed, there was so little difference between the two NBC networks that one announcer, a bit worse for wear, began his broadcast:

"Good morning: this is either the Red or the Blue network."

In the days when ABC had difficulty competing with the much richer CBS and NBC for entertainment programming, it signed Aaron Spelling (with his partner Leonard Goldberg) to become exclusive suppliers to the network. Spelling became such a force in the network's destiny that some people began to call ABC Aaron's Broadcasting Company.

During the fifties and sixties ABC seemed forever doomed to be the third and last network in terms of ratings, and it became the regular butt of industry jokes.

"Do you know how to end the Korean (or Vietnam) war?" Answer: "Put it on ABC and it will be over in 13 weeks."

Following the Symbionese Liberation Army's kidnaping of Patty Hearst, a joke went: "Why can't the FBI find Patty Hearst?" Answer: "She's hiding on ABC."

Rock 'Round the Clock

NBC headquarters has been known as "30 Rock" because of its address at Rockefeller Plaza. CBS headquarters on West 57th in New York has usually been called Black Rock, some say because of the color of the granite building, others that it was named after the 1955 MGM movie, *Bad Day at Black Rock*. ABC, for decades a poor relative, moved in 1965 to Sixth Avenue, near to CBS, which began jokes that it occupied the crate in which the CBS building was packed. Following the rock theme, ABC's executive offices have also been called Little Rock (when Southerner Tom Moore was its president), Schlock Rock, and Hard Rock. When ABC finally became the number-one network in the mid-1970s, its building became known as Hot Rock.

Foul-up

In the 1970s, when Fred Silverman was the whiz-kid programmer of American television entertainment, a friend of his asked whether he intended to spend Yom Kippur at home. The executive inquired on what day the holiday fell.

"Wednesday," the friend said.

"Wednesday?" Silverman exploded. "You mean they've scheduled Yom Kippur opposite *Charlie's Angels?*"

Fair Warning

The mannerisms of the title characters in the American television sitcom *Laverne and Shirley*, depicting two single girls, are so different from what passes for normal behavior outside the United States that the television station in Bangkok,

Thailand, would only show it with a Siamese caption at the beginning of each episode: "The two women depicted in this series live in an insane asylum."

Wrong Way

Brandon Tartikoff, the youngest president and chairman of NBC before he departed for Paramount Pictures, studied literature at Yale under Robert Penn Warren. Blessed even then with a commercial imagination, Tartikoff once offered the critique of a short story by Tolstoy that one of the characters should have been a car. It was then that his Pulitzer prize–winning professor suggested that Tartikoff might turn his attention to television.

Not the Best or the Brightest

Linda Elstad tells a story of a documentary filmmaker who had once worked as a writer of television drama in Hollywood but now was employed in the much less lucrative area of educational television. Why had he switched, Elstad wanted to know, since she was a television writer herself. This is how she remembers his answer.

Way back when a great deal more work was available in episodic television for free lancers, this writer got a call from his agent, who said he had arranged an appointment for him at MGM in half an hour to pitch his ideas. Since the writer lived in the San Fernando Valley, at least half an hour from the studio, he needed to get into the shower and then his car at once.

"What kind of show is it?" he asked quickly.

"It's a detective show," said the agent, "and I don't know anything more about it. But I told the producers that you had lots of great ideas."

A few minutes later the writer was heading toward Culver City and en route trying to come up with just one great idea. He recalled an old adage in the business: "When in difficulty, steal from the greatest." That is how his mind turned eventually to the greatest one of them all: Shakespeare. He thought of *Macbeth*—no—then *Hamlet* . . . —no—but when he came to *Othello*—the light went on: bingo! He would think up a story about jealousy, but he could also improve on Shakespeare. Having been a police reporter, the writer had been troubled by Desdemona making speeches and taking all that time to die. He knew that death by strangling was fatal and final: you're either dead, or you ain't. Such poetic license should not be allowed in episodic television, he thought. So, he began his pitch to the producers with an opening scene showing a man strangling a woman. After this teaser and the introductory titles, the woman was no longer to be found on the sofa, which would later suggest to the detective hero investigating her murder that the man seen at the beginning could not have killed her, unless he had somehow managed to remove her body.

This opening gimmick hooked the producers right away, and one of them exclaimed:

"This is just great, it's a fantastic concept, you've got a deal. But before you go off and write it, I want our secretary to take the idea down on paper."

The producer called his secretary and began to dictate: "Opening scene. We see a woman lying on the sofa, and then this guy comes in, picks up a lamp, and hits her on the head . . ."

Not long afterward, the writer who told me this story, found a compelling need to become an educator, and that is how he began making documentaries.

Services Rendered

Following the quiz-show scandals of 1959, Walter Lippmann, the dean of American journalists, attacked the networks for pandering to the lowest tastes and proposed the establishment of a non-profit, public broadcasting system. It was soon after this that Fred Friendly, then the chief producer of documentaries at CBS, approached Lippmann for a wide-ranging interview, which became an annual institution. But despite his enormously enhanced influence, Lippmann was of the old school and did not feel comfortable with the hoopla of television. One day, when he reproached Friendly, half kidding, for dragging him into broadcasting, the producer reminded him:

"Walter, if it weren't for television and me you'd still be stuck in the classified pages of the *Washington Post* and the *Herald Tribune*."

Lippmann, normally a modest man, wanted the network man to know that the service rendered had not been one-sided.

"And I made you respectable, young man."

They Like to Watch

The debate on the harmful effects of television, especially on children, has been raging almost since TV first invaded our homes. One writer disturbed by the impact of the medium was the late Jerzy Kosinski, the author of *Being There*, the story of a simple gardener who becomes a media guru and presidential candidate. Everything he knows is derived from and defined by the images emanating from "the box." Kosinski was not writing only fiction based on his imagination. In the early seventies, he had conducted personal experiments with public-school children between the ages of nine and fourteen, one of which he described to Susan Stamberg on National Public Radio:

"The most horrifying experiment was when I was being televised telling a group of children an interesting story. I had arranged that, in the middle of the story, I was to be assaulted by a man breaking into the classroom. Now, as I was telling the story to the children, they had a choice of either watching me 'in person,' as I sat in front of them, or watching me on two giant television screens we'd placed on either side of the room. As long as I was alone and nothing bad was happening, they looked directly at me. The minute the assailant came and started to punch me and I started to bleed, most children in the classroom turned toward the TV screens. Suddenly, I was without any help from anyone. They refused to look at me. They chose instead to watch the assault as a television show."

Make My Point

In 1978, an American advertising man by the name of Jerry Mander caused a minor sensation in intellectual circles with his book, *Four Arguments for the Elimination of Television*. Naturally, he was much in demand for television interviews. One producer called and asked Mander if he could sum up in a few words why he wanted to get rid of television. Mander replied that he could not, and that is why he took a hundred thousand words to express himself. Then the producer, who clearly had no intention of reading the book, asked what his main points were.

"One of the main points," Mander obliged, "is that television can only deal with main points."

Bibliography

Books Cited and Consulted

Abbe, Patience, Richard, and Johnny. *Of All Places!* New York: Frederick A. Stokes, 1937.

Abramson, Albert. *The History of Television, 1880 to 1941.* Jefferson, N.C.: McFarland & Co., 1987.

Ace, Goodman. *The Book of Little Knowledge—More Than You Want to Know About Television.* New York: Simon & Schuster, 1955.

Adams, Edie, and Robert Windeler. *Sing a Pretty Song . . .* New York: Morrow, 1990.

Adler, Bill, and Jeffrey Feinman. *Mel Brooks—The Irreverent Funny Man.* New York: Playboy Press, 1976.

Allen, Fred. *Much Ado About Me.* Boston: Little, Brown, 1956.

Allen, Steve. *Bigger Than a Breadbox.* Garden City, N.Y.: Doubleday, 1967.

————. *Funny People.* Briarcliff Manor, N.Y.: Stein & Day, 1981.

Allen, Thomas B., F. Clifton Berry, and Norman Polmar. *CNN—War in the Gulf.* Atlanta: Turner Publishing, 1991.

Andrews, Bart. *Lucy & Ricky & Fred & Ethel—The Story of "I Love Lucy."* New York: Dutton, 1976.

Anon. *Stranger Than Fiction—By "The Voice of Experience."* New York: Dodd, Mead, 1934.

———. *Washington Broadcast—By the Man at the Microphone.* Garden City, N.Y.: Doubleday, Doran, 1944.

Arlen, Michael J. *Living Room War.* New York: Viking, 1969.

———. *Thirty Seconds.* New York: Farrar, Straus & Giroux, 1980.

Auerbach, Arnold M. *Funny Men Don't Laugh.* Garden City, N.Y.: Doubleday, 1965.

Auletta, Ken: *Three Blind Mice—How the TV Networks Lost Their Way.* New York: Random House, 1991.

Autry, Gene, with Mickey Herskowitz. *Back in the Saddle Again.* Garden City, N.Y.: Doubleday, 1978.

Aylesworth, Thomas G. *Great Moments of Television.* New York: Exeter Books, 1987.

Bacon, James. *Hollywood Is a Four-Letter Town.* Chicago: Contemporary Books, 1976.

———. *Made in Hollywood.* Chicago: Contemporary Books, 1977.

Baker, Richard. *Here Is the News.* London: Leslie Frewin, 1966.

Barnouw, Erik. *A Tower of Babel.* New York: Oxford University Press, 1966.

———. *The Golden Web.* New York: Oxford University Press, 1968.

———. *The Image Empire.* New York: Oxford University Press, 1970.

———. *Tube of Plenty.* New York: Oxford University Press, 1990.

Barris, Alex. *The Pierce-Arrow Showroom Is Leaking—An Insider's View of the CBC.* Toronto: The Ryerson Press, 1969.

———. *Front Page Challenge.* Toronto: Canadian Broadcasting Corporation, 1981.

Barsley, Michael. *Behind the Screen—Stories from the Backroom of the BBC.* London: André Deutsch, 1957.

Benny, Joan and Jack. *Sunday Nights at Seven—The Jack Benny Story.* New York: Warner Books, 1990.

Bergen, Candice. *Knock Wood.* New York: Simon & Schuster, 1984.

Berle, Milton, with Haskel Frankel. *An Autobiography.* New York: Delacorte Press, 1974.

Bestlaffs of the Year. New York: Harper & Brothers, 1927.

Bilby, Kenneth. *The General—David Sarnoff and the Rise of the Communications Industry.* New York: Harper & Row, 1986.

Bishop, Jim. *The Golden Ham—A Candid Biography of Jackie Gleason.* New York: Simon & Schuster, 1956.

Black, Peter. *The Biggest Aspidistra in the World—A Personal Celebration of 50 Years of the BBC.* London: BBC, 1972.

Blair, Gwenda. *Almost Golden—Jessica Savitch and the Selling of Television News.* New York: Simon & Schuster, 1988.

Bloomfield, Roderick (ed.). *Heard in the Wings.* London: Stanley Paul, 1971.

Bohannon, John. *Kickers: All the News That Didn't Fit.* New York: Ballantine, 1988.

Borns, Betsy. *Comic Lives—Inside the World of American Stand-Up Comedy.* New York: Simon & Schuster, 1987.

Boyer, Peter J. *Who Killed CBS?* New York: Random House, 1988.

Broun, Heywood Hale. *Tumultuous Merriment.* New York: Richard Marek, 1979.

Brown, Les. *Televi$ion—The Business Behind the Box.* New York: Harcourt Brace Jovanovich, 1971.

Burnett, Carol. *One More Time.* New York: Random House, 1986.

Burns, George, with David Fisher. *All My Best Friends.* New York: Putnam, 1989.

Burrows, Abe. *Honest Abe.* Boston: Little, Brown, 1980.

Burstein, Patricia. *Farrah—An Unauthorized Biography of Farrah Fawcett-Majors.* New York: New American Library, 1977.

Buxton, Frank, and Bill Owen. *The Big Broadcast 1920–1950.* New York: Viking, 1972.

Byrne, Gay. *To Whom It Concerns.* Dublin: Torc Books, 1972.

Caesar, Sid, with Bill Davidson. *Where Have I Been?* New York: Crown, 1982.

Cahn, William. *Good Night, Mrs. Calabash—The Secret of Jimmy Durante.* New York: Duell, Sloan & Pearce, 1963.

Cannon, Lou. *President Reagan—The Role of a Lifetime.* New York: Simon & Schuster, 1991.

Cantril, Hadley. *The Invasion from Mars—A Study in the Psychology of Panic.* Princeton, N.J.: Princeton University Press, 1966.

Carpozi, George, Jr., *Vince Edwards—Television's Ben Casey.* New York: Belmont Books, 1962.

———. *The Carol Burnett Story.* New York: Warner Paperback Library, 1975.

Carr, William H. A. *What Is Jack Paar Really Like?* New York: Lancer Books, 1962.

Carroll, Carroll. *None of Your Business—Or My Life with J. Walter Thompson.* New York: Cowles Book Company, 1970.

Castleman, Harry, and Walter J. Podrazik. *Watching TV—Four Decades of American Television.* New York: McGraw-Hill, 1982.

Cavett, Dick, and Christopher Porterfield. *Cavett.* New York: Harcourt Brace Jovanovich, 1974.

Celebrity Research Group. *The Bedside Book of Celebrity Gossip.* New York: Crown, 1984.

Cerf, Bennett. *Anything for a Laugh.* New York: Grosset & Dunlap, 1946.

———. *Laughter Incorporated.* Garden City: Garden City Books, 1950.

———. *The Laugh's on Me.* New York: Doubleday, 1959.

———. *Laugh Day.* New York: Doubleday, 1965.

———. *The Sound of Laughter.* New York: Doubleday, 1970.

Chandler, Charlotte. *Hello, I Must Be Going—Groucho & His Friends.* Garden City, N.Y.: Doubleday, 1978.

Christensen, Mark, and Cameron Stauth. *The Sweeps—Behind the Scenes in Network TV.* New York: Morrow, 1984.

Clark, Ian G. *Canada, Broadcasting, and Me.* Richmond, British Columbia: Darkwood Holding Company, 1990.

Cockerell, Michael. *Live from Number 10—The Inside Story of Prime Ministers and Television.* London: Faber & Faber, 1989.

Collins, Pan. *It Started on the Late, Late Show.* Dublin: Ward River Press, 1981.

Corwin, Norman. *Trivializing America.* Secaucus, N.J.: Lyle Stuart, 1983.

Cosell, Howard, with Mickey Herskowitz. *Cosell.* New York: Playboy Press, 1973.

————, with Peter Bonventre. *I Never Played the Game.* New York: Morrow, 1985.

Craft, Christine. *An Anchorwoman's Story.* Santa Barbara, Calif.: Capra Press, 1986.

Crosby, John. *Out of the Blue.* New York: Simon & Schuster, 1952.

————. *With Love and Loathing.* New York: McGraw-Hill, 1963.

Crowther, Bruce, and Mike Pinfold. *Bring Me Laughter.* London: Columbus Books, 1987.

Cuthbert, Margaret (ed.). *Adventure in Radio.* Howell, Soskin, Publishers, 1945.

Dallas, Paul V. *Dallas in Wonderland—The Pacifica Approach to Free Radio.* Los Angeles: self-published, 1967.

Daly, Marsha. *Steve Martin—An Unauthorized Biography.* New York: New American Library, 1980.

Davis, Gerry. *The Today Show—An Anecdotal History.* New York: Morrow, 1987.

Day, Holman. *The Ship of Joy—Captain Hugh Barrett Dobbs, Commander.* San Francisco: n.p., 1931.

Day, Sir Robin. *Television—A Personal Report.* London: Hutchinson, 1961.

————. *Grand Inquisitor.* London: Weidenfeld and Nicolson, 1989.

De Cordova, Fred. *Johnny Came Lately.* New York: Simon & Schuster, 1988.

Della Femina, Jerry, edited by Charles Sopkin. *From Those Wonderful Folks Who Gave You Pearl Harbor.* New York: Simon & Schuster, 1970.

DeMille, Cecil B., and Donald Hayne (ed.). *The Autobiography of Cecil B. DeMille.* Englewood Cliffs, N.J.: Prentice-Hall, 1959.

D'Essen, Lorrain. *Kangaroos in the Kitchen—The Story of Animal Talent Scouts.* New York: David McKay, 1959.

Dizard, Wilson P. *Television—A World View.* Syracuse, N.Y.: Syracuse University Press, 1966.

Dobbs, Fred C. *The Golden Age of B.S.* Toronto: Gage, 1976.

Donaldson, Sam. *Hold On, Mr. President!* New York: Random House, 1987.

Downey, Morton, Jr., with William Hoffer. *Mort! Mort! Mort! No Place to Hide.* New York: Delacorte Press, 1982.

Downs, Hugh. *Yours Truly.* New York: Holt, Rinehart & Winston, 1960.

Dreher, Carl. *Sarnoff—An American Success.* New York: New York Times Book Co., 1977.

Duncan, Peter. *"In Town Tonight."* London: Werner Laurie, 1951.

Dunn, Clive. *Permission to Speak—An Autobiography.* London: Century, 1986.

Dunning, John. *Tune in Yesterday—The Ultimate Encyclopedia of Old-Time Radio.* Englewood Cliffs, N.J.: Prentice-Hall, 1976.

Dunphy, Don. *Don Dunphy at Ringside.* New York: Henry Holt, 1988.

Edmonson, Madeleine, and David Rounds. *From Mary Noble to Mary Hartman— The Complete Soap Opera Book.* Briarcliff Manor, N.Y.: Stein & Day, 1976.

Eells, George, and Stanley Musgrove. *Mae West.* New York: Morrow, 1982.

Efron, Edith. *The News Twisters.* New York: Manor Books, 1972.

Eliot, Mark. *American Television—The Official Art of the Artificial*. Garden City, N.Y.: Anchor Press/Doubleday, 1981.

Ellerby, Linda. *"And So It Goes"—Adventures in Television*. New York: Putnam, 1986.

Elliott, Bob, and Ray Goulding. *From Approximately Coast to Coast . . . It's the Bob and Ray Show*. New York: Atheneum, 1983.

Epstein, Edward Jay. *News from Nowhere*. New York: Random House, 1973.

Esslin, Martin. *Mediations—Essays on Brecht, Beckett, and the Media*. Baton Rouge: Louisiana State University Press, 1980.

———. *The Age of Television*. Stanford, Calif.: Stanford Alumni Association, 1981.

Fabe, Maxene. *TV Game Shows!* Garden City, N.Y.: Doubleday, 1979.

Fang, Irving E. *Those Radio Commentators!* Ames, Iowa: Iowa State University Press. 1977.

Fates, Gil. *What's My Line?* Englewood Cliffs, N.J.: Prentice-Hall, 1978.

Fedo, Michael. *The Man from Lake Wobegon*. New York: St. Martin's Press, 1987.

Fein, Irving A. *Jack Benny—An Intimate Biography*. New York: Putnam, 1976.

Ferguson, Max. *And Now . . . Here's Max*. Toronto: McGraw-Hill, 1967.

Feuer, Jane, Paul Kerr, and Tise Vahimagi. *MTM—"Quality Television."* London: British Film Institute, 1984.

Fitzgibbon, Constantine. *The Life of Dylan Thomas*. Boston: Little, Brown, 1965.

Floherty, John J. *Behind the Microphone*. Philadelphia: J. B. Lippincott, 1944.

Fowler, Gene, and Bill Crawford. *Border Radio*. New York: Limelight Editions, 1990.

Fox, Stephen. *The Mirror Makers*. New York: Morrow, 1984.

Foxx, Redd, and Norma Miller. *The Redd Foxx Encyclopaedia of Black Humor*. Pasadena, Calif.: Ward Ritchie Press, 1975.

Freeman, Don. *Eyes as Big as Cantaloupes*. San Diego, Calif.: Joyce Press, 1978.

———. *In a Flea's Navel*. San Diego: A. S. Barnes, 1980.

Freeman, Rita Grade. *My Fabulous Brothers—The Story of the Grade Family*. London: W. H. Allen, 1982.

Friendly, Fred W. *Due to Circumstances Beyond Our Control . . .* New York: Random House, 1967.

Frum, Barbara. *As It Happened*. Toronto: McClelland & Stewart, 1976.

Fry, William F., Jr., and Melanie Allen. *Make 'Em Laugh—Life Studies of Comedy Writers*. Palo Alto, Calif.: Science and Behavior Books, 1975.

Galanoy, Terry. *Down the Tube*. New York: Pinnacle Books, 1972.

Gates, Gary Paul. *Air Time—The Inside Story of CBS News*. New York: Harper & Row, 1978.

Gaver, Jack, and Dave Stanley. *There's Laughter in the Air!* New York: Greenberg, 1945.

Gilbert, Douglas. *Floyd Gibbons—Knight of the Air*. New York: Robert M. McBride and Co., 1930.

Goldenson, Leonard H., with Martin J. Wolf. *Beating the Odds*. New York: Scribner, 1991.

Goldmark, Peter, with Lee Edson. *Maverick Inventor—My Turbulent Years at CBS.* New York: Saturday Review Press/E. P. Dutton, 1973.

Gordon, George N., and Irving A. Falk. *On the Spot Reporting—Radio Records History.* New York: Julian Messner, 1967.

Granlund, Nils Thor, with Sid Feder and Ralph Hancock. *Blondes, Brunettes, and Bullets.* New York: David McKay, 1957.

Green, Abel, and Joe Laurie, Jr. *Show Biz—From Vaude to Video as Seen by VARI-ETY.* New York: Holt, 1951.

Green, Abel (ed.). *The Spice of Variety.* New York: Henry Holt, 1952.

Griffin, Merv, with Peter Barsochini. *Merv—An Autobiography.* New York: Simon & Schuster, 1980.

Grisewood, Frederick. *The World Goes By.* London: Secker & Warburg, 1952.

Gross, Ben. *I Looked & I Listened.* New Rochelle, N.Y.: Arlington House, 1970.

Grossman, Gary H. *Saturday Morning TV.* New York: Arlington House, 1987.

Guest, Edgar A. *It Can Be Done.* Chicago: The Reilly & Lee Co., 1938.

Gunther, John. *Taken at the Flood—The Story of Albert D. Lasker.* New York: Harper & Brothers, 1960.

Hall, Monty, and Bill Libby. *Emcee Monty Hall.* New York: Grosset & Dunlap, 1973.

Hall, Willis and Bob Monkhouse. *The A-Z of Television.* London: Pelham Books, 1971.

Halle, Kay. *The Irrepressible Churchill.* New York: Facts on File, 1985.

Harding, Gilbert. *Along My Line.* London: Putnam & Co., 1953.

Harmon, Jim. *The Great Radio Heroes.* Garden City, N.Y.: Doubleday, 1967.

———. *The Great Radio Comedians.* Garden City, N.Y.: Doubleday, 1970.

Harris, Credo Fitch. *Microphone Memoirs of the Horse and Buggy Days of Radio.* Indianapolis: Bobbs-Merrill, 1937.

Harvey, Paul, Jr. (ed.). *Paul Harvey's For What It's Worth.* New York: Bantam Books, 1991.

Hecht, Andrew. *Hollywood Merry-Go-Round.* New York: Grosset & Dunlap, 1947.

Hewitt, Don. *Minute by Minute.* New York: Random House, 1985.

Hill, Doug, and Jeff Weingrad. *Saturday Night.* New York: Morrow, 1986.

Hill, George H. *Airwaves to the Soul—The Influence and Growth of Religious Broadcasting in America.* Saratoga, Calif.: R & E Publishers, 1983.

Hope, Bob, as told to Pete Martin. *Have Tux, Will Travel—Bob Hope's Own Story.* New York: Simon & Schuster, 1954.

Hoyt, Ken, and Frances Spatz Keighton. *Drunk Before Noon—The Behind-the-Scenes Story of the Washington Press Corps.* Englewood Cliffs, N.J.: Prentice-Hall, 1979.

Huntley, Chet. *The Generous Years.* New York: Random House, 1968.

Husing, Ted, with Cy Rice. *My Eyes Are in My Heart.* New York: Bernard Geis Associates, 1959.

Johnson, Lesley. *The Unseen Voice—A Cultural Study of Early Australian Radio.* London: Routledge, 1988.

Johnson, William O., Jr. *Super Spectator and the Electric Lilliputians*. Boston: Little, Brown, 1971.

Johnston, Brian. *Chatterboxes—My Friends the Commentators*. London: Methuen, 1983.

Julian, Joseph. *This Was Radio—A Personal Memoir*. New York: Viking Press, 1975.

Kaltenborn, H. V. *Fifty Fabulous Years*. New York: Putnam, 1950.

Kavanagh, Ted. *Tommy Handley*. London: Hodder and Stoughton, 1946.

Keeley, Joseph. *The Left-Leaning Antenna—Political Bias in Television*. New York: Arlington House, 1971.

Kilgallen, Dorothy. *Girl Around the World*. Philadelphia: David McKay, 1936.

King, Larry, with Emily Yoffe. *Larry King*. New York: Simon & Schuster, 1982.

Kirck, Harvey, with Wade Rowland. *Nobody Calls Me Mr. Kirck*. Toronto: Collins, 1985.

Kitman, Marvin. *The Marvin Kitman TV Show*. New York: Outerbridge & Lazard, 1972.

Klavan, Gene. *We Die at Dawn*. Garden City, N.Y.: Doubleday, 1964.

Koop, Theodore F. *Weapon of Silence*. Chicago: University of Chicago Press, 1946.

LaGuardia, Robert. *From Ma Perkins to Mary Hartman*. New York: Ballantine, 1977.

Lamparski, Richard. *Whatever Became Of . . . ?* New York: Crown, 1967.

Land, Myrick. *The Fine Art of Literary Mayhem*. San Francisco: Lexicos, 1983.

Laporte, Marcel. *Les Memoires de Radiolo*. Paris: Bernard Grasset, 1925.

Latham, Aaron. *Perfect Pieces*. New York: Arbor House, 1987.

Lax, Eric. *Woody Allen—A Biography*. New York: Knopf, 1991.

Leaming, Barbara. *Orson Welles*. New York: Viking Penguin, 1985.

Ledyard, Gleason H. *Sky Waves—The Incredible Far East Broadcasting Company Story*. Chicago: Moody Press, 1968.

Levant, Oscar. *The Memoirs of an Amnesiac*. New York: Putnam, 1965.

Lewis, Thomas S. W. *Empire of the Air—The Men Who Made Radio*. New York: Edward Burlingame Books, 1991.

Lunden, Joan, with Ardy Friedberg. *Good Morning, I'm Joan Lunden*. New York: Putnam, 1986.

Lusty, Robert. *Bound to Be Read*. Garden City, N.Y.: Doubleday, 1976.

Lyman, Darryl. *The Jewish Comedy Catalog*. Middle Village, N.Y.: Jonathan David, 1989.

Machlin, Milt. *The Gossip Wars—An Exposé of the Scandal Era*. Self-published, 1981.

Madsen, Axel. *60 Minutes—The Power and the Politics of America's Most Popular TV News Show*. New York: Dodd, Mead, 1984.

Manio, Jack de. *To Auntie with Love*. London: Hutchinson, 1967.

———. *Life Begins Too Early*. London: Hutchinson, 1970.

Mansell, Gerard. *Let Truth Be Told—50 Years of BBC External Broadcasting*. London: Weidenfeld and Nicolson, 1982.

Marx, Groucho, with Hector Arce. *The Secret Word Is Groucho*. New York, Putnam, 1976.

Maskell, Dan, with John Barrett. *Oh, I Say!* London: Collins, 1988.

Mathews, Joseph J. *Reporting the Wars*. Minneapolis: University of Minnesota Press, 1957.

Matusow, Barbara. *The Evening Stars*. New York: Houghton Mifflin, 1983.

Maud, Ralph (ed.). *On the Air with Dylan Thomas—The Broadcasts*. New York: New Directions, 1992.

Mayer, Martin. *About Television*. New York: Harper & Row, 1972.

McBride, Mary Margaret. *Out of the Air*. Garden City, N.Y.: Doubleday, 1960.

McCann, Irving G. *Case History of the Smear by CBS of Conservatives*. Washington, D.C: self-published, 1966.

McCarthy, Joe (ed.). *Fred Allen's Letters*. Garden City, N.Y.: Doubleday, 1965.

McCrohan, Donna. *The Second City*. New York: Putnam, 1987.

———. *Prime Time, Our Time*. Rocklin, Calif.: Prima Publishing, 1990.

McKenzie, Michael. *Backstage at Saturday Night Live!* New York: Scholastic Book Services, 1980.

McLendon, Winzola, and Scottie Smith. *Don't Quote Me!—Washington News-women and the Power Society*. New York: Dutton, 1970.

McNamee, Graham, with Robert Gordon Anderson. *You're on the Air*. New York: Harper & Brothers, 1926.

Metz, Robert. *CBS—Reflections in a Bloodshot Eye*. New York: Playboy Press, 1975.

———. *The Today Show*. New York: Playboy Press, 1977.

Miall, Leonard. *Richard Dimbleby, Broadcaster—By His Colleagues*. London: British Broadcasting Corporation, 1966.

Michael, Paul, and James Robert Parish. *The Emmy Awards—A Pictorial History*. New York: Crown, 1970.

Midgley, Leslie. *How Many Words Do You Want?* New York: Carol Publishing Group, 1989.

Miller, Bill, as told to John McCarty. *You're on Open Line!—Inside the Wacky World of Late-Night Talk Radio*. Brattleboro, Vt.: The Stephen Greene Press, 1978.

Millner, Cork. *Santa Barbara Celebrities—Conversations from the American Riviera*. Santa Barbara, Calif.: Santa Barbara Press, 1986.

Mingo, Jack. *The Official Couch Potato Handbook*. Santa Barbara, Calif.: Capra Press, 1983.

Mitz, Rick. *The Great TV Sitcom Book*. New York: Richard Marek Publishers, 1980.

Moger, Art. *Some of My Best Friends Are People*. Boston: Challenge Press, 1964.

Monaco, James. *Media Culture*. New York: Dell Publishing, 1978.

More, Kenneth. *More or Less—An Autobiography*. London: Hodder & Stoughton, 1978.

Morella, Joe, and Edward Z. Epstein. *Lucy*. New York: Lyle Stuart, 1973.

Mortimer, John. *In Character—Interviews with Some of the Most Influential and Remarkable Men and Women of Our Time*. London: Allen Lane, 1983.

Muggeridge, Malcolm. *Tread Softly for You Tread on My Jokes*. London: Collins/Fontana, 1968.

Mulryan, Peter. *Radio Radio—The Story of Independent, Local, Community and Pirate Radio in Ireland*. Dublin: Borderline Publications, 1988.

Murray, Kathryn. *Family Laugh Lines*. Englewood Cliffs, N.J.: Prentice-Hall, 1966.

Murrow, Edward R., and Fred W. Friendly (eds.). *See It Now*. New York: Simon & Schuster, 1955.

Nash, Alanna. *Golden Girl—The Story of Jessica Savitch*. New York: Dutton, 1988.

New Statesman Profiles. London: Readers Union, 1958.

Norden, Denis, Sybil Harper, and Norma Gilbert. *Coming to You Live!—Behind-the-Screen Memories of Forties and Fifties Television*. London: Methuen, 1985.

Oppenheimer, Jerry. *Barbara Walters—An Unauthorized Biography*. New York: St. Martin's, 1990.

Osgood, Charles. *The Osgood Files*. New York: Putnam, 1991.

Paisner, Daniel. *The Imperfect Mirror—Inside Stories of Television Newswomen*. New York: Morrow, 1989.

Palmer, Myles. *Woody Allen*. London and New York: Proteus, 1980.

Partridge, Marianne (ed.). *Rolling Stone Visits Saturday Night Live*. Garden City, N.Y.: Doubleday, 1979.

Paul, Eugene. *The Hungry Eye—An Inside Look at TV*. New York: Ballantine Books, 1962.

Perry, George. *Life of Python*. Boston: Little, Brown, 1983.

Pertwee, Michael. *Name Dropping*. London: Leslie Frewin, 1974.

Pixton, Ralph. *On the Line*. Hong Kong: Zebra Books, 1978.

Playfair, Guy Lion. *The Evil Eye—The Unacceptable Face of Television*. London: Jonathan Cape, 1990.

Plomley, Roy. *Days Seemed Longer*. London: Methuen, 1980.

Poindexter, Ray. *Golden Throats and Silver Tongues*. Conway, Ark.: River Road Press, 1978.

Powers, Ron. *The Newscasters*. New York: St. Martin's, 1977.

Quinn, Sally. *We're Going to Make You a Star*. New York: Simon & Schuster, 1975.

Randall, Tony, with Michael Mindlin. *Which Reminds Me*. New York: Delacorte Press, 1989.

Raphael, Sally Jessy, with Pam Proctor. *Sally—Unconventional Success*. New York: 1990.

Rather, Dan, with Mickey Herskowitz. *The Camera Never Blinks*. New York: Morrow, 1977.

Reader's Digest Treasury of Wit & Humor. Pleasantville, N.Y.: The Reader's Digest Association, 1958.

Reagan, Ronald, with Richard G. Hubler. *Where's the Rest of Me?* New York: Duell, Sloan & Pearce, 1965.

Reasoner, Harry. *Before the Colors Fade*. New York: Knopf, 1981.

Reid, Colin. *Action Stations—A History of Broadcasting House*. London: Robson Books, 1987.

Rivers, Joan, with Richard Meryman. *Enter Talking*. New York: Delacorte Press, 1986.

Rogers, Lynne. *The Love of Their Lives*. New York: Dell, 1979.

Rovin, Jeff. *TV Babylon*. New York: New American Library, 1984.

Rubin, Benny. *Come Backstage with Me*. Bowling Green, Ohio: Bowling Green University Popular Press, n.d.

Saerchinger, Cesar. *Hello America!—Radio Adventures in Europe*. Boston: Houghton Mifflin, 1938.

Sanders, Coyne Steve. *Rainbow's End—The Judy Garland Show*. New York: Morrow, 1990.

Sanford, Herb. *Ladies and Gentlemen, The Garry Moore Show*. Briarcliff Manor, N.Y.: Stein & Day, 1976.

Sarnoff, David. *Looking Ahead—The Papers of David Sarnoff*. New York: McGraw-Hill, 1968.

Schafer, Kermit. *Bloopers, Bloopers, Bloopers*. New York: Bell Publishing, 1973.

Schechter, A. A., with Edward Anthony. *I Live on Air*. New York: Frederick A. Stokes, 1941.

Scherz, Ede. *A Rádio Humora*. Budapest: self-published, 1931.

Schoenbrun, David. *On and Off the Air*. New York: Dutton, 1989.

Sears, William. *God Loves Laughter*. London: George Ronald, 1960.

Seldes, Gilbert. *The Great Audience*. New York: Viking Press, 1950.

Seuling, Barbara. *You Can't Show Kids in Underwear and Other Little-Known Facts about Television*. Garden City, N.Y.: Doubleday, 1982.

Sevareid, Eric. *Not So Wild a Dream*. New York: Knopf, 1946.

Shales, Tom. *On the Air!* New York: Summit Books, 1982.

Shanks, Bob. *The Cool Fire—How to Make It in Television*. New York: Norton, 1976.

Shulman, Arthur, and Roger Youman. *How Sweet It Was*. New York: Bonanza Books, 1966.

Silvers, Phil, with Robert Saffron. *This Laugh Is on Me—The Phil Silvers Story*. Englewood Cliffs, N.J.: Prentice-Hall, 1973.

Singer, Mark. *Mr. Personality—Profiles and Talk Pieces*. New York: Knopf, 1988.

60 Minutes Verbatim—Season XII. New York: Arno Press/CBS News, 1980.

Smith, Richard, and Edward Decter. *Oops!—The Complete Book of Bloopers*. New York: Rutledge Press, 1981.

Smith, Ron. *Sweethearts of '60s TV*. New York: St. Martin's, 1989.

Smith, Sally Bedell. *In All His Glory—The Life of William S. Paley*. New York: Simon & Schuster, 1990.

Soares, Manuela. *The Soap Opera Book*. New York: Harmony Books, 1978.

Speece, Wynn, with M. Jill Karolevitz. *The Best of the Neighbor Lady*. Dakota Homestead Publishers, 1987.

Stallings, Penny, with Howard Mandelbaum. *Flesh and Fantasy*. New York: St. Martin's Press, 1978.

Stamberg, Susan. *Susan Stamberg's All Things Considered Book*. New York: Pantheon, 1982.

Stansky, Peter. *On Nineteen Eighty-Four.* Stanford, Calif.: Stanford Alumni Association, 1983.

Stefoff, Rebecca. *Mary Tyler Moore: The Woman Behind the Smile.* New York: New American Library, 1986.

Stein, Ben. *The View from Sunset Boulevard—America as Brought to You by the People Who Make Television.* New York: Basic Books, 1979.

Stein, M. L. *When Presidents Meet the Press.* New York: Julian Messner, 1969.

Stern, Bill, with Oscar Fraley. *The Taste of Ashes.* New York: Henry Holt, 1959.

Stuart, Lyle. *The Secret Life of Walter Winchell.* Boar's Head Books, 1953.

Sugar, Bert Randolph. *"The Thrill of Victory"—The Inside Story of ABC Sports.* New York: Hawthorn Books, 1978.

Sweet, Jeffrey. *Something Wonderful Right Away.* New York: Avon Books, 1978.

Talbot, Godfrey. *Ten Seconds from Now—A Broadcaster's Story.* London: Hutchinson, 1973.

Taylor, Robert Lewis. *Doctor, Lawyer, Merchant, Chief.* Garden City, N.Y.: Doubleday, 1948.

Teague, Bob. *Live and Off-Color: News Biz.* New York: A & W Publishers, 1982.

Teichmann, Howard. *George S. Kaufman: An Intimate Portrait.* New York: Atheneum, 1972.

———. *Smart Aleck—The Wit, World and Life of Alexander Woollcott.* New York: Morrow, 1976.

Tennis, Craig. *Johnny Tonight!* New York: Simon & Schuster, 1980.

Thomas, Lowell. *Good Evening Everybody.* New York: Morrow, 1976.

Tormé, Mel. *The Other Side of the Rainbow—With Judy Garland on the Dawn Patrol.* New York: Morrow, 1970.

Treadwell, Bill. *Head, Heart and Heel.* New York: Mayfair Books, 1958.

Trueman, Peter. *Smoke & Mirrors—The Inside Story of Television News in Canada.* Toronto: McClelland & Stewart, 1980.

Tynan, Kathleen. *The Life of Kenneth Tynan.* New York: Morrow, 1987.

Tynan, Kenneth. *Show People—Profiles in Entertainment.* New York: Simon & Schuster, 1979.

Vallee, Rudy. *Kisses & Tells.* Canoga Park, Calif.: Major Books, 1976.

Vig, György. *Halló! Itt Rádio Humor!* Budapest: self-published, n.d.

Wapner, Joseph A. *A View from the Bench.* New York: Simon & Schuster, 1987.

Warner, Henry Edward. *Uncle Ed and His Dream Children.* Baltimore: n.p., 1929.

Waters, James F. *The Court of Missing Heirs.* New York: Modern Age Books, 1941.

Webb, Richard, and Teet Carle. *The Laugh's on Hollywood.* Santa Monica, Calif.: Roundtable Publishing, 1985.

Weiner, Ed. *Let's Go to Press—A Biography of Walter Winchell.* New York: Putnam, 1955.

Wertheim, Arthur Frank. *Radio Comedy.* New York: Oxford University Press, 1979.

Wheen, Francis. *Television.* London: Century Publishing, 1985.

Whelan, Kenneth. *How the Golden Age of Television Turned My Hair to Silver.* New York: Walker & Co., 1973.

Whittemore, Hank. *CNN—The Inside Story*. Boston: Little, Brown, 1990.

Whittinghill, Dick, with Don Page. *Did You Whittinghill This Morning?* Chicago: Henry Regnery Co., 1976.

Wilk, Max. *The Golden Age of Television—Notes from the Survivors*. New York: Delacorte Press, 1976.

Williams, Huntington. *Beyond Control—ABC and the Fate of the Networks*. New York: Atheneum, 1989.

Williams, Kenneth. *Acid Drops*. London: J.M. Dent, 1980.

Wilson, Earl. *Pikes Peek or Bust*. Garden City, N.Y.: Doubleday, 1946.

———. *Let 'em Eat Cheesecake*. Garden City, N.Y.: Doubleday, 1949.

———. *The Show Business Nobody Knows*. New York: The Cowles Book Company, 1971.

———. *Show Business Laid Bare*. New York: Putnam, 1974.

———. *Hot Times—True Tales of Hollywood and Broadway*. Chicago: Contemporary Books, 1984.

Wogan, Terry. *Wogan on Wogan*. London: Penguin, 1988.

Wood, Rob (ed.). *The Book of Blunders*. Kansas City, Mo.: Hallmark Editions, 1974.

Wyand, Paul. *Useless If Delayed—Adventures in Putting History on Film*. London: Harrap, 1959.

Yacowar, Maurice. *Method in Madness—The Comic Art of Mel Brooks*. New York: St. Martin's Press, 1981.

Young, Jimmy. *Jimmy Young*. London: Michael Joseph, 1982.

Zicree, Marc Scott. *The Twilight Zone Companion*. New York: Bantam, 1982.

Zolotow, Maurice. *No People Like Show People*. New York: Random House, 1951.

Newspapers and Magazines

Atlantic, *Broadcasting*, *Canadian Theatre Review*, *Channels*, *The Daily News*, *Entertainment Weekly*, *Esquire*, *Hollywood Reporter*, *Life*, *The Listener*, *The Los Angeles Times*, *Media History Digest*, *New Statesman*, *Newsweek*, *The New York Times*, *The New Yorker*, *Nielsen Researcher*, *Old Time Radio Digest*, *Playboy*, *Premiere*, *SPERDVAC Radiogram*, *Stage Magazine*, *Take One*, *Time*, *TV Guide*, *TV—Könyv*, *Vanity Fair*, *Variety*, *Video Age International*, *Video Times*, *View*.

Index of Broadcasters, Personalities, and Programs